Owls Aren't Wise & Bats Aren't Blind

TN

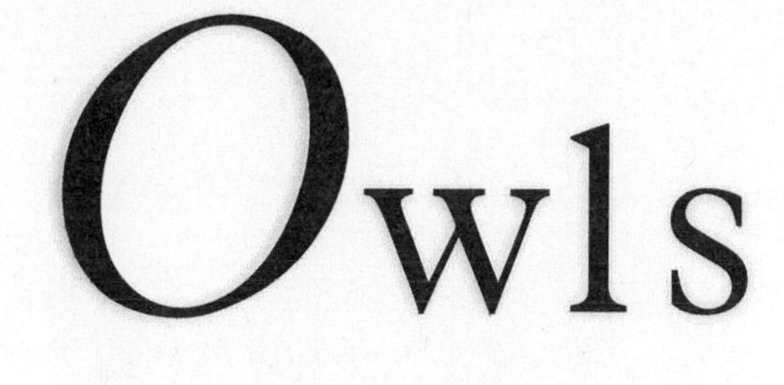

Owls Aren't Wise & Bats Aren't Blind

A Naturalist Debunks Our Favorite Fallacies About Wildlife

WARNER SHEDD

Illustrated by Trudy Nicholson

THREE RIVERS PRESS
NEW YORK

Published by Three Rivers Press, New York, New York.
Member of the Crown Publishing Group.

Random House, Inc. New York, Toronto, London, Sydney, Auckland

www.randomhouse.com

Originally published in hardcover by Harmony Books in 2000.

Printed in the United States of America

Designed by Lenny Henderson

Library of Congress Cataloging-in-Publication Data
Shedd, Warner, 1934–
Owls aren't wise and bats aren't blind : a naturalist debunks our favorite fallacies about wildlife / by Warner Shedd.
Includes bibliographical references.
1. Animal behavior. 2. Animal intelligence. I. Title.
QL751 .S56 2000
591.5—dc21 99-58301

ISBN 978-0-609-80797-2

20 19 18 17 16 15

First Paperback Edition

To Edie, my wife and most trusted critic,

and to the rest of my family for their

continued interest in and support of this project

Contents

acknowledgments

It would take an enormously long list to thank individually the many, many wildlife biologists and researchers who so generously shared their knowledge and time with me; therefore, a general expression of my gratitude will have to suffice. These busy individuals were unfailingly courteous, amazingly patient, and often delightfully friendly and encouraging while answering what must sometimes have seemed like a deluge of questions. These people are the true wildlife experts, and I salute them.

Special thanks go to my literary agent, Linda Roghaar. With seemingly infinite patience, she guided me through the process of creating this book, beginning with her invaluable advice and assistance in preparing a book proposal that she was subsequently able to market. Always good-humored and willing to answer my questions or help me in other ways too numerous to mention, Linda is the epitome of what I believe a good literary agent should be.

Trudy Nicholson, who illustrated this book, also deserves profound thanks. Her wonderful artistic skill, her almost obsessive desire for accuracy in even the smallest details, and her ability to work long hours, day after day, in order to meet deadlines were an enormous asset.

Thanks are also due to my editor, Patricia Gift, and her assistant, Kristen Wolfe. They, too, were very patient with my lack of knowledge about the production of a book by a large publishing house and did a great deal to make the process run more smoothly.

Owls

Aren't Wise

&

Bats

Aren't Blind

introduction

THE PATH THAT LED TO THIS BOOK BEGAN WITH AN INSTINCTIVE FASCINATION WITH WILDLIFE, EVIDENT EVEN WHEN I WAS VERY YOUNG. While I was growing up, my parents, my two sisters, and I lived on my maternal grandparents' dairy farm in Vermont's Champlain Valley. My grandfather, who was a business school graduate and machinist before he became a farmer, was quite artistic, and my favorite object in our home was his large, framed pencil drawing of a family of gray squirrels in a huge oak tree. I used to gaze at it for long periods of time, as it was a remarkably good portrait for an untrained artist. To my immense and continuing pleasure, it now graces the wall of my office, hanging just beneath the long-barreled muzzle-loader that was my grandfather's squirrel gun when he was a boy.

A large drawer in my grandfather's desk contained a very nice set of antlers from a whitetail buck that he had shot many years before. These also fascinated me, and from time to time I would ask to look at them. Then one obliging adult or another would open the drawer, take them out, and let me handle them for a minute or two before putting them away again.

One day when I was five or six, my grandfather announced that the next day was May 1 (the traditional opening day of the trout season in those days), and that he was going to take me fishing. I was immensely excited, of course, since I had never before been fishing.

The next day Grandpa and I set out for the creek that formed one boundary of our farm, a half-mile or more from the house. Across the meadow we went, down the lane, through pastures and meadows, and eventually beneath the tall, lovely pines that grew in a broad swath of woods between pasture and creek. Finally we reached our destination—a very large boulder beneath which the current swirled and carved a nice little pool in its shady recesses.

The creek, where it bordered our farm, was too far downstream to harbor trout. No doubt it had at some distant time, when Native Americans hunted and fished along its banks. Now, however, it flowed through enough miles of open land to allow the sun to heat the water beyond the range tolerated by trout. There was still good trout fishing not many miles upstream, I later learned, but not where we were about to fish.

All of this made not the slightest difference to me. The pool beneath the rock contained numerous chub and dace—members of the minnow tribe—that eagerly grabbed the worms with which my grandfather helped me bait my hook. Each fish was a new adventure, and I loved every moment of the experience.

At the same time, I came under the spell of the creek itself. True, its waters weren't cold enough for trout, those most demanding aristocrats of the fish world, but it ran clean and clear. The sight and sound of the moving water as it sparkled, rushed, gurgled, and murmured its way over and around the numerous rocks in its path captivated me utterly—a captivation that continues to this day. Like the Mole in *The Wind in the Willows,* I had entered into the joy of running water!

Our return home was triumphant. I was praised to the skies for the fine catch of dace and chub that I brought home. These were duly cleaned, dredged in flour, sautéed in butter, and presented for my delectation. I thought that I had never eaten anything finer!

School followed—a one-room school a quarter-mile from our home. There was one teacher for eight grades, so of necessity we students spent much time working on our own. Once our lessons were done, we could either ask for drawing paper and amuse ourselves drawing or coloring, or we could read. Sometimes I chose to draw pictures, and these were invariably of birds and animals. Most of the time, however, I read.

The school library, if it could be called that, was minuscule, consisting of a little revolving stand about three feet high that contained a handful of volumes. I did manage, at an early age, to acquire a nodding acquaintance with such things as Greek and Norse mythology from its scanty resources. My greatest pleasure, however, came from the bookmobile that stopped by every two or three weeks. From its manifold treasures I almost always selected books about wildlife. I especially recall several of Ernest Thompson Seton's books, as well as a little volume written by a naturalist, whose name I'm not sure of now, that contained color plates of several of the animals discussed in it. I was particularly delighted by the author's reference to the skunk, at one point, as "Sir Mephitis."

Our home library was also very limited, but I sometimes received books about wildlife for Christmas or birthdays. One, titled *Horns and Antlers* as I recall, educated me considerably about deer, elk, antelope, mountain goats, and the like. We also had a very out-of-date encyclopedia that further enlightened me about the natural world.

At age nine I developed a strong urge to hunt; after some very careful checking, I was deemed sufficiently responsible to go afield with a little single-shot .22 rifle in quest of the wary gray squirrels that inhabited the woodlands on and near the farm. I was indeed responsible enough to hunt at that age, for I was almost obsessive about safety when I had a gun in my hand.

As my stalking and shooting skills improved, I brought home numerous squirrels for the pot. Actually, we didn't put them into regular stews; instead, they were usually sautéed or fricasseed with sour cream to make a delicious gravy to accompany the sweet, tender meat.

At about that time, I read a book by Gene Stratton Porter—I believe it was called *Moths of the Limberlost*—that greatly expanded my interests in the natural world. Soon I was collecting our native giant silkworm moths—cecropia, luna, Io, and their kin. Over the next three or four years I managed to amass a fairly creditable display of these gorgeous but seldom seen creatures.

More or less simultaneously, I discovered a very old book called *Boys at Chequasset, or a Little Leaven*. This was about a rather careless, irresponsible boy whose parents moved into the country. There he became interested in collecting birds' eggs, and this pursuit eventually led him to reform his irresponsible ways. What I now recognize as the book's tedious and excessively Victorian morality managed to escape me at that age, but the part about collecting birds' eggs and displaying them properly fired my imagination. Soon I, too, was collecting eggs, albeit in a very responsible manner. I only removed a single egg from a nest, and then only if the bird was in the early stages of egg-laying, so that she would be sure to lay a replacement egg.

The woods and streams had become my natural habitat, whether I was fishing, hunting, searching for birds' eggs and moths, or simply enjoying the dim coolness of the forest. My interests also expanded far beyond those creatures that were my quarry. The sight of a pileated woodpecker, a white egret (a rare visitor), or a bittern was enough to make that day a memorable one, regardless of anything else.

When the time for college arrived, natural resource management was a natural career path for me. Actually I chose forest management rather than wildlife management, for several reasons that aren't germane here. Six years later, armed with a B.S. and an M.S., I began an eight-year stint as a forester for the Vermont Department of Forests and Parks.

During that time it became apparent that we had better become concerned about increasingly severe environmental problems. Rachel Carson's

Silent Spring illuminated the horrors that were being caused by many pesticides, such as DDT, which caused eggshells to become so thin that they broke prematurely in the nest, endangering eagles, ospreys, and other species. Many rivers were open sewers, and one, the Cuyahoga, actually caught on fire! Lake Erie, into which the Cuyahoga flowed, was considered a nearly dead body of water. I became an active volunteer with the National Wildlife Federation, and conservation became both my vocation and my avocation.

My interests had now extended well beyond forest management. When a position as area resource specialist for the Vermont Extension Service became available, I jumped at the opportunity. This work greatly broadened my interests and experience, while I also continued my volunteer work within the framework of the National Wildlife Federation.

After three years of extension service work, I was asked to become the Eastern (later New England) Regional Executive for the National Wildlife Federation. This major career shift placed me squarely in the thick of environmental advocacy. It was exhilarating work that dealt at various levels with international, national, and regional issues.

At the national level were issues concerning air and water pollution, pesticides, habitat preservation, and obtaining passage of the Alaska Lands Act. At the state and regional levels were such things as the successful battles to halt the ill-advised Dickey-Lincoln and Big A dams in Maine, and to create several new wilderness areas in the Green Mountain National Forest in Vermont.

Throughout my entire professional career, I had the opportunity to associate closely with the people whom I consider to be the true wildlife experts—professional wildlife biologists and researchers. Many of these became good friends, and I shamelessly picked their brains and peppered them with questions about wildlife. Their uniformly patient, thorough responses added enormously to my store of knowledge about a great many native species of wildlife.

Gradually I became increasingly aware of—and more and more disturbed by—the myths and fallacies about wildlife that I heard with great frequency. Some were excusable misinterpretations or misunderstandings by people far removed from the land and wildlife. Others were ludicrous, yet were accepted and repeated uncritically by people who believed them simply because they had heard them since childhood.

Finally, when I embarked on a new career as a writer, I resolved, if possible, to correct a great many of these erroneous ideas about wildlife. The result was this book. My foremost hope is that you will find this volume, so stunningly illustrated by Trudy Nicholson, thoroughly enjoyable. I've tried to provide accurate scientific information, in an accessible and engaging format. Some scientists will no doubt think I haven't explained some things in sufficient detail to be totally accurate; I beg their indulgence on the grounds that this book is not intended for scientists, but for the lay reader who wants to know more about the fascinating lives of animals.

My second hope, obviously, is to debunk much of the incorrect information that encrusts many of our common species of wildlife. The more we understand the true nature of wildlife and how it lives, the more we come to appreciate and value it.

Finally, I hope this book conveys the endless joy and fascination that I find in wildlife. May you find your life enriched, as mine has been, by these wild fellow inhabitants of our fragile planet.

1

Nature's Engineer: The Beaver

Myths

- Beavers pack mud with their tails.
- Beavers can fell trees in a desired direction.
- Beavers always build dams.
- "Bank beavers" are a different species.
- Beavers eat fish.
- If a beaver loses its mate, it won't mate again.
- "Nuisance" beavers should be moved to another location.

THE SINGLE MOST BASIC AND IMPORTANT FACT ABOUT THE BEAVER *(CASTOR CANADENSIS)* IS THIS: NEXT TO HUMANS, THE BEAVER IS UNDOUBTEDLY THE MOST PROFICIENT OF NORTH AMERICAN INHABITANTS AT MANIPULATING THE ENVIRONMENT FOR ITS OWN BENEFIT. For several hundred thousand years we humans have been able to harness fire and manufacture a variety of tools, ranging from primitive stone and bone implements to modern steam shovels, to assist us in modifying our surroundings. Beavers, however, have become legendary dam builders by using only instinct and the tools with which nature has endowed them.

Beavers have the ability to modify their immediate environment quite radically, even when they have very little to work with. Give a pair of beavers a reasonably flat area, a scant trickle of water, a modest supply of trees or shrubs, and they can quickly create a dam—large or small, as the situation demands—a pond, and a dwelling secure from predators. In the process they quite unintentionally create a great deal of valuable marsh or wetland habitat for many other species.

Perhaps because beavers are such intriguing creatures, with astonishing engineering and construction ability, they're one of our most misunderstood mammals. Through the years, they've become encrusted with a number of major misconceptions that often obscure their true nature. For example, one of the most common folk myths is that beavers pack mud, or carry mud, with their tails. They don't. How this notion gained such currency is anyone's guess, but it simply isn't true: although a beaver's tail has many uses, packing mud isn't one of them. Whenever beavers need to carry or pack mud while constructing their dams or houses, they use their front paws.

That fact fails to diminish in the slightest the importance of the beaver's tail. Broad, flat, hairless, and scaly, it's shaped like the blade of an old-fashioned canoe paddle. In fact the familiar and still-popular canoe paddle design is referred to as a beavertail paddle.

The tail's unique conformation makes it suitable for so many purposes that it serves its owner as a virtual tool kit. For openers, the tail is the beaver's alarm signal. Anyone who's spent much time around a beaver pond, especially at dusk, is familiar with the loud *ka-WHOP* that resounds across the still waters. When a beaver detects human presence—or any other form of potential danger—it warns the other members of its colony by slamming its tail forcefully onto the surface of the water. The result is a mighty splash similar to what one would expect if a fifty-pound rock were heaved into the water!

Of necessity, beavers spend much of their time swimming, and the wide tail acts as both rudder and diving plane while the beaver moves to and fro, up and down, within its watery domain. A beaver's tail is also very fatty, and this fat can act as an emergency food supply in time of famine. And because it's hairless, the tail functions as a heat exchange mechanism during periods when its owner is exceptionally active. It even serves as a brace or prop while the beaver sits up and gnaws during the process of felling trees. Clearly, the beaver's tail is the very model of a utilitarian appendage—but the one thing it doesn't do is pack mud!

Another common myth is that beavers, like human loggers, can fell trees in the desired direction. The truth is far less glamorous. The beaver is a rodent, with incisors that act as big, sharp chisels. When a beaver decides to cut down a tree, it usually gnaws a V-shaped notch completely round the tree, slanting up from the bottom and down from the top to meet at the apex of the cut.

The beaver continues to work around the tree, all the while deepening and enlarging the notch. Eventually there's too little wood left in the middle to support the tree, and it falls, willy-nilly, in any old direction. Far from knowing or planning the direction of fall, beavers are occasionally killed when the trees they're felling accidentally land on them. Sometimes a beaver will gnaw through a tree from only one side, but that choice assuredly isn't dictated by any intention of felling the tree toward a particular spot.

It's fascinating to speculate on the many ways in which misconceptions about wildlife begin. Some are fairly easy to fathom, others less so. We like to be charitable and think that people are honest, but some individuals make up tall tales and pass them off as the truth, or just plain lie for reasons best known only to themselves. A case in point is the conversation I had with one man about beavers.

He assured me with a perfectly straight face that he had personally watched beavers packing mud with their tails. Further, he said that he had seen a beaver fell a hardwood tree over a foot in diameter in less than five minutes. Beavers can remove sizable chips with those big front teeth, but they certainly aren't that fast. Such a feat would tax most people with a sharp ax and saw! The fact is that a beaver requires at least two or three nights of hard work, and sometimes more, to take down a single large tree.

Beavers are indisputably dam builders *par excellence,* justly famous for this ability. They live exclusively in colonies, and construction of a dam is the first step in founding a colony for a newly mated pair of young beavers. There's a good reason for this. A high percentage of our North American beavers live where winter is severe and the ice becomes very thick. Beavers store their winter food supply underwater, and they require water deep enough that neither the underwater entrances to their home nor their food supply will be encased by ice.

This is purely instinctive behavior that has nothing to do with reasoning. Beavers evolved under harsh conditions that necessitated water depth sufficient for winter survival. Now they'll mindlessly build dams wherever they possibly can, even if there's no apparent reason for it. Further, they'll continue to raise the height of a dam year after year, oblivious to the fact that the depth of their pond was more than sufficient from the beginning. Even in large lakes and ponds, beavers will attempt to dam the outlet, despite water depth many times what they require for survival.

It's incorrect to assume that beavers always build dams, however. Beavers often live in wide, deep streams and rivers that are too difficult a challenge even for their formidable dam-building skills. Under such conditions, beavers tunnel up into the riverbank to create a home. Beavers living in this fashion are often referred to as "bank beavers," and old-timers frequently claimed that they were a different species. They aren't, but they do seem to have learned that they can't dam that particular stream, and have somehow managed to override their age-old compulsion to build a dam.

Dam construction itself is a fascinating process. The beavers begin by laying down brush, with the butts pointed downstream. Mud and rocks are brought to the site to anchor and coat the brush, and various materials, from pieces of driftwood to old bottles, are gradually incorporated into the dam as it grows higher and higher. Mud is constantly packed in among the sticks, brush, and other material, and the upstream surface of the dam is coated and recoated with mud.

The whole affair is really a beaver version of concrete. Just as we use cement, weak by itself, to bind gravel and iron rods into extremely strong concrete, beavers use mud as their cement, with brush, sticks, and other materials as the reinforcing medium. The end result is a structure that's strong and nearly impervious to water, except for the overflow through the brush and sticks at the very top of the dam.

A beaver dam is such a fascinating edifice that it's worth inspecting at close range, both for its ingenious construction and for the variety of materials often incorporated into it. Given the near ubiquity of beavers today, most people shouldn't find it difficult to locate a beaver dam that they can scrutinize at leisure and thereby gain a much greater appreciation of this clever rodent's abilities.

In addition to the physical construction of the dam itself, beavers have some related talents. For one thing, they seem to have an uncanny eye for selecting dam sites that offer a high ratio of pond acreage to size of dam. From the beaver's point of view, the perfect dam site combines a short span between high banks with a wide, flat area just upstream. A minimum amount of construction in this sort of location will yield a large, deep pond.

Perhaps an even more remarkable instinct causes beavers to vary a dam's conformation according to conditions. If the current is slow or the span short, the dam is constructed more or less in a straight line. However, longer dams where the current is strong are curved, with the convex side pointing upstream. Of course, this just happens to be the way in which humans engineer big dams!

After building their dam to a satisfactory height, the beaver pair next fabricate a dome-shaped house called a lodge. Like the dam, the lodge is made of a mixture of sticks and mud, although it's constructed with far less care. The beavers simply heap everything up, starting from the bottom of the pond and extending the mass for several feet above water level. Then, beginning at a safe depth underwater, they gnaw and dig their way upward until they're above water level. There they scoop and gnaw out a roomy chamber.

The size of the lodge depends on the number of beavers in the colony. (More about colony size and composition later.) Lodges six feet or more above water level aren't uncommon, and in exceptionally old colonies—forty to fifty years old—lodges have been found as high as twelve to fifteen feet above water level.

The interior of the lodge may consist of a single, all-purpose chamber, or it may have a feeding chamber and a connected resting chamber. A good deal of room is required to house a number of beavers, and the interior of an extremely large lodge may be capable of seating several people at once!

Once the lodge is finished, it provides the beavers with great security. Just as the steel rods in reinforced concrete add a great deal of strength, the high ratio of sticks to mud in a beaver lodge makes it highly resistant to attack. Even in summer, with relatively thin walls surrounding the above-water chamber, it's very difficult for a predator to break into the lodge, and in that event the beavers can simply exit into the safety of their pond. In winter, when the mud freezes, the lodge becomes as hard as iron, impregnable even to the largest, fiercest predators.

Although a beaver lodge is admirably suited for ensuring the survival of its occupants, we shouldn't regard it as the equivalent of a human dwelling. Whenever he gives a lecture about beavers, Thomas Decker, a biologist with the Vermont Fish and Wildlife Department, likes to point out that a beaver lodge doesn't fit our notion of a nice, cozy home. The interior is hot and humid in the summer, smelly at all seasons of the year, and infested with a variety of undesirable trespassers. Hordes of insects often abound inside a lodge, water snakes like to crawl into it, and otters may enter from time to time; the latter, incidentally, will try to kill young beavers up to the age of a month or two. Clearly, this is no palace by human standards!

It's widely and erroneously believed that beavers include fish in their diet. However, beavers are entirely vegetarian. In summer, they mainly consume a variety of aquatic plants, algae, and some grasses and ferns. Their principal winter food supply is the inner bark of various species of trees and shrubs. Aspen, willow, alder, and maple are favorites, although they will eat a variety of other hardwood species, including ash and birch. Beavers normally eschew the bark of evergreens as food, but will turn to it if forced to by a shortage of better foods.

When winter draws near, the beavers cut quantities of tree branches and shrubs and bury the butt ends in the mud next to their lodge. As soon as the pond becomes icebound, the beavers retire to the safety of their lodge; when they want food, they swim out, cut sections of branches, bring them back to the lodge, and munch off the bark. The peeled sticks are carried outside to float up under the ice; otherwise their debris would soon fill the lodge. Although bark represents their primary winter food supply, beavers also supplement their diet with the thick, tuberous rootstocks of pond lilies, at least in those ponds where these plants grow.

Like a number of other wild creatures, beavers mate for life, but that doesn't mean what most people seem to think it does. The common perception of lifetime mating in the wild is that the death of one partner relegates the survivor to a single existence for the remainder of its life. That's not the case at all. Like many humans, a wild animal that normally mates for life will, after the death of its mate, seek another.

As previously noted, beavers live in a colony that consists exclusively of a single beaver family. Each colony begins with a mated pair of young beavers. After the pair have constructed dam and lodge, and laid down a winter's food supply, they breed during the winter. After a gestation of about ninety days, the young, called kits, are born in the spring. There are usually three to five kits in a litter, and occasionally more.

The young remain with the parents through the winter, and the following spring the colony is enlarged by a second litter of kits. The colony now contains the parents plus the offspring from two successive years. All of the beavers in a colony share the work of maintaining the dam and lodge, and cutting and storing the winter food supply.

Although a number of myths have grown up concerning beavers, the phrase "busy as a beaver" most assuredly isn't one of them. Beavers labor intensively during the warm months, preparing for the long winter to come. Constant diligence and great effort are required to ensure an adequate and secure habitat, as well as an ample winter food supply. Only when the pond finally becomes icebound can the beavers relax and enjoy the fruits of their labors.

Big changes occur during the colony's third spring. As another litter of kits is born, the young of two years before either leave the colony voluntarily or are driven out by their parents. These "dispossessed" beavers must now seek new territory, find mates, and begin new colonies. Stark evidence of this springtime diaspora is often visible in the form of dead beavers killed by automobiles as they try to cross highways.

Thereafter, for the life of the colony, a new set of kits will be born each spring, and the two-year-olds will either set out on their own or be driven out.

Although this system may seem harsh, it's necessary for the colony's survival. Without this annual dispersal, a beaver colony would soon become impossibly overpopulated, to the detriment of all.

The manner in which the dispersed two-year-olds find mates is an interesting one, worthy of a bit of exploration. The beaver's genus name, *Castor,* comes to us through Latin from *Kastor,* the Greek name for the beaver. Eventually, *castor* even became a term for a beaver hat during the heyday of that article of apparel.

Of greater relevance in this instance is that beavers of both sexes secrete an oily, pungent substance that's also known as castor. This rather malodorous liquid is produced by a pair of glands in the inguinal region, close to the anus. Beavers are highly territorial and have a well-developed sense of smell, so as soon as a young beaver finds a suitable site for a dam, it marks the boundaries of its territory with what are called *castor mounds.*

Castor mounds are piles of mud and debris carried by the beaver with its front paws. These heaps are normally placed close to the water, and the beaver deposits castor on them. Such mounds can grow into sizable structures, especially where the territories of two or more colonies intersect. In that situation, a beaver from one colony will try to cover the scent of another colony by heaping more material on the mound and then depositing its own castor. As this "castor duel" continues, the mounds can sometimes grow to waist height on a human!

Because of its excellent sense of smell, a two-year-old beaver scouring the countryside for a home will soon locate and investigate castor mounds. If the mounds' originator proves to be an unmated beaver of the opposite sex, a new colony has its start.

Beavers are obviously highly aquatic mammals, and they have a number of wonderful evolutionary adaptations that make it easy for them to function in and under the water. When it comes to moving about in the water, beavers are extremely well equipped. In addition to the nearly all-purpose tail, the beaver has large, fully webbed hind feet to provide efficient propulsion. These can thrust the big animal through the water with ease, even when the front feet are occupied with carrying mud or rocks.

Beavers can also stay underwater for an extraordinarily long time—at least fifteen minutes at a stretch—and they accomplish this feat with lungs of normal size. How is this possible? First, their heartbeat drops to half its usual rate when the beavers are underwater, thereby reducing their metabolic rate and their oxygen consumption. Second, they can also utilize most of the oxygen in their lungs—an astonishing 75 percent versus 15 percent in humans!

Nor does working underwater pose problems for the beaver. The instant a beaver submerges, certain reflex actions take place. A transparent "third eyelid," called a nictitating membrane, covers each eye to allow good underwater vision. Simultaneously, special membranes close off the ears and nostrils. Meanwhile, a similar membrane blocks the throat and permits the beaver to gnaw wood or gather aquatic plants beneath the surface without swallowing water.

To top off the long list of adaptations, the beaver also has a double fur coat. The outer coat consists of long, glossy guard hairs, while the inner coat is so dense that water doesn't even reach the beaver's skin. With this sort of protection, the beaver has no difficulty swimming and diving in the most frigid water.

The beaver's wonderful fur coat, so necessary for its survival, was once the cause of its near demise. When explorers and settlers came to North America, they found beavers in enormous quantities. As they sent beaver pelts, often obtained by trading with Native Americans, back to Europe, beaver hats and coats grew steadily in popularity. Soon beaver garments became all the rage in Europe, and the North American fur trade grew to enormous proportions. Indeed, it's been said, without too much exaggeration, that the trade in beaver furs built North America.

Literature from the late sixteenth century until well into the nineteenth century reflects the immense and continuing popularity of beaver hats. Oliver Wendell Holmes, for example, describing the well-dressed man, adjured his readers to "Have a good hat; the secret of your looks / Lives with a beaver in Canadian brooks."

An idea of the immense demand for beaver fur can be gleaned from the fact that the Hudson's Bay Company alone sold over three *million* beaver pelts during a twenty-five-year period in the second half of the nineteenth century. As beavers were trapped relentlessly in the absence of any conservation laws, forests were simultaneously felled to make room for farmland. The combination of unregulated trapping and vanishing habitat proved deadly. By 1900, beavers had become scarce virtually everywhere and were extirpated throughout much of their range. Then the pendulum slowly began to swing.

With the advent of modern wildlife management techniques, laws to protect beavers were passed and strictly enforced. Beavers were also live-trapped in areas where they were still plentiful, and restocked in regions where they had been eliminated. For example, in 1921, six beavers were trapped in New

York State and released in southern Vermont; eleven years later, beavers were reintroduced from Maine into northeastern Vermont.

No species, however rigorously protected, can survive without adequate habitat, and here the ways of humans, once so inimical to the beaver's survival, began to favor it. Concurrent with protective laws and the reintroduction of beavers, there was a slow but steady abandonment of agricultural land. This land soon reverted to the forest habitat ideal for beavers. Slowly at first, and then with increasing rapidity, the beavers returned to their old haunts.

The restoration of the beaver has had many beneficial effects. Countless wetlands, once largely dried up, now dot the landscape, helping to hold back high water and recharging underground aquifers. Even more important, perhaps, is the vast amount of habitat that beavers create for other species of wildlife. Indeed, in terms of benefits conferred on other wildlife species, beavers have to be awarded the palm as our single most important wild mammal.

Biologists credit beavers for their important role in the remarkable resurgence of moose populations in areas such as northern New England. Moose thrive on—indeed, require—the summer nourishment provided by aquatic vegetation in shallow ponds, and many beaver ponds are ideal for the growth of those plants. Further, moose need water where they can stay cool in the heat of summer and escape, at least temporarily, the swarms of biting insects so common throughout their range.

Moose are by no means the only beneficiaries of the beaver's work. Muskrats, otters, and other mammals utilize beaver flowages. Many waterfowl species nest on the ground along the shores of beaver ponds and raise their broods on the ponds. Cavity-nesting birds, including wood ducks and goldeneyes, find nest sites in standing dead trees killed by the flooding from beaver flowages. Herons, bitterns, egrets, and other wading birds find fertile hunting grounds along the shallow margins of these myriad bodies of water.

Other forms of life benefit from the beavers as well. Fish often thrive in beaver ponds, at least for a few years after the pond's creation. Frogs, turtles, salamanders, water snakes, insects, and other aquatic creatures occupy these new wetlands in large numbers. Indeed, a whole host of creatures benefits from the beaver and its wetlands.

After a period of years, a beaver colony's supply of food and building materials begins to diminish. As food becomes scarcer, the beavers resort to various strategies to stretch out the life of the colony. Normally, beavers don't go much more than one hundred yards from their home ponds in search of food, but with declining food supplies they may increase that distance. They will also dig canals leading away from a pond so that they can transport food and building materials to the pond by water. As a last resort, they'll even begin to exploit normally shunned food sources, such as the bark of evergreens.

Almost inevitably, however, supplies finally run out, and the beavers depart. However, their departure doesn't by any means signal the end of the beaver's usefulness in creating habitat for other species. Without constant attention, the dam soon deteriorates and gradually disappears, leaving a wide, muddy expanse. Fertilized by beaver droppings, this mud flat quickly revegetates, first with grasses and other herbaceous plants, and later with shrubs and tree seedlings.

The former pond is now called a beaver meadow—a very different type of habitat from the beaver pond, but nonetheless a valuable one. Deer, moose, and other creatures come here to graze, browse, and otherwise feed on this rich new food source. Gradually, though, the shrubs and trees grow large enough to provide a food supply for beavers once again. Then beavers return, and the cycle repeats itself, as it has for thousands of years.

Not all is well with the resurgence of the beaver, however. At first it was greeted joyfully by most people. As they became more and more numerous, however, conflicts between beavers and their works and humans began to mount. Now beavers, once a favorite of nearly everyone, are so numerous that they've become pariahs in the eyes of many people. As a result, they've created discord not only between beavers and humans, but also between beaver lovers and beaver haters. What created this remarkable shift in the attitude of so many individuals toward the big rodents?

The beaver's reproductive potential is a good place to begin seeking answers. Once beavers are well established, they can multiply very rapidly. They begin to breed at age two or three and can easily live for twelve to fifteen years, and sometimes as much as twenty. Thus, even allowing for normal attrition among their offspring, a single pair of beavers can produce a very large number of survivors in the course of a lifetime.

Abetting this reproductive capacity is the relative lack of predation. The beaver is a large animal. It happens to be the world's second largest rodent, topped only by the South American capybara. Adult beavers average forty to fifty pounds, and exceptionally large specimens can reach seventy, eighty, even ninety pounds. In fact, the largest beaver ever recorded actually exceeded one hundred pounds. (Although large by present-day standards, the beaver is a midget compared with one of its ancestors: a sort of giant beaver named *Trogonotherium* lived about a million years ago and was seven feet long!)

Any animal that weighs fifty pounds or more, and is armed with a set of powerful chisel teeth, is a formidable opponent for all but the biggest and strongest predators. Prior to the advent of European settlers, the only major North American predators of beavers were Native Americans, wolves, and cougars. Much of the predation by the latter two was probably carried out at the time when the two-year-old beavers dispersed; traveling overland or along shallow streams much of the time, these itinerant young beavers were highly vulnerable to large predators.

Nowadays, wolves and cougars control beavers only where enough wild country remains to support those big predators. However, it's unreasonable and impractical to expect wolves and cougars to return to populated areas. Bears, bobcats, and coyotes will kill an occasional young beaver, and otters prey on baby beavers now and then, but their predation is too sporadic to have much overall effect on such a prolific animal.

That leaves humans as the only major restraint—other than starvation and disease—on beaver populations. For a few decades, trapping held beaver populations at a reasonable level, more or less in balance with habitat. This system began to come unglued, however, when felt hats—the best ones, manufactured from the beaver's underfur—went out of style in the 1960s. Prices for beaver pelts declined and, as a consequence, so did human predation on beavers. With reduced predation, beavers multiplied and became an increasingly serious nuisance in many areas.

When I was a small child growing up in Vermont, my parents took me to a museum in which a Vermont map displayed the location of all the state's known beaver colonies, indicated by a handful of widely scattered red dots. Today a similar map would be almost totally red, so abundant are the state's thousands of beaver colonies!

As more and more young beavers seek places to establish a colony, they gradually occupy every available nook and cranny. In fact, I've seen numerous dams and ponds so tiny that they can have no possible value to the beavers except, out of sheer desperation, to satisfy their dam-building compulsion. This constant, widespread dam construction often brings the builders into sharp conflict with humans. For example, beavers love to plug culverts, thereby flooding highways. The beavers don't do it to be malicious, of course:

they're acting purely from instinct. To them, a culvert simply seems like a very narrow constriction in a stream—a perfect place to build a dam. Malicious or not, however, the road is still flooded, and the beavers have to go.

Getting rid of beavers in this situation is not simple. Tearing a big hole in their dam (no easy feat, incidentally, as many have discovered) is a futile gesture. Beavers are thoroughly industrious creatures, and the sound of running water—other than that trickling through and over the very top of their dam—is anathema to them. Consequently, they'll plug a large gap in their dam in a single night. Worse yet, beavers are as persistent as they are hardworking, and they'll continue to repair the damage night after night until human patience is exhausted.

Dynamite may then become the weapon of choice, but even blowing a beaver dam to hell and gone won't work unless it's done so late in the year that the beavers have no chance to rebuild before winter. Otherwise the dam is back in a trice, the beavers are again in business, and flooding remains a problem. If the dam is totally destroyed late in the year, however, the beavers are doomed to starve or fall to predators as soon as winter arrives.

So-called "beaver cheaters"—long, perforated pipes or wood-and-screen structures inserted through the dam—sometimes work. With their upstream ends near the bottom of the pond and far enough above the dam, these devices are sometimes able to confuse the beavers and prevent them from discovering the source of the leakage. At other times, though, the dam builders, with their uncanny ability to detect current, will locate and plug the offending structures.

Another serious problem is that beavers fail to distinguish between fruit and shade trees and wild trees. Homeowners are understandably upset when this happens, yet the beavers are often forced to utilize every inch of available habitat—including cultivated trees and shrubs. This, of course, is small solace to the person whose prize fruit trees or treasured shade trees suddenly resemble the aftermath of George Washington's mythical "Father, I cannot tell a lie" episode.

A neighbor of mine described just such an incident. Several years ago he lived on the edge of a large, marshy area. Beavers soon appropriated this spot and built a dam and lodge. There were several homes bordering this new beaver pond, and for a time everyone was delighted by the new residents. Then the problems began.

Each year the beavers raised the dam and the water level a bit, and after four or five years the water began creeping into septic-system leachfields. Moreover, the beavers began to run out of food and sought whatever they could find around their pond. What they could find happened to include the fruit trees in this man's backyard. Formerly very pro-beaver, his opinion, he admitted, underwent a rapid and dramatic shift at that point. "I like beavers," he commented dryly, "but I think I like apple trees more."

Unsuccessful efforts were made either to evict the beavers by peaceable means, or to lower the level of their impoundment. The dam was torn out repeatedly, and was promptly restored by the beavers on each occasion. Then supposedly beaverproof outlet structures were installed, but the clever and industrious rodents managed to plug them all.

Finally an adjoining homeowner, whose septic system was seriously impaired by the beaver flowage, obtained a permit from the state to kill the beavers. This caused a neighborhood squabble, with some opposed to killing the beavers (mainly those who hadn't suffered beaver damage), and others in favor of their summary execution (mainly those who *had* suffered damage).

In the midst of this turmoil, the beavers themselves solved the problem. Constantly seeking new sources of food and building supplies, they began to cross a busy paved highway adjoining their pond. Beavers weren't designed by nature for crossing highways safely, and humans in automobiles soon became predators, albeit unintentionally, of these particular specimens.

Similar damage recently made headlines when beavers gnawed down some of the famous Washington, D.C., cherry trees near the Jefferson Memorial. The beavers had taken up residence in the Tidal Basin between the Jefferson Memorial and the Washington Monument, and in characteristic beaver fashion began to search for materials for dam and lodge, as well as food. Park Service officials managed to live-trap them and move them to another location, thereby at least temporarily forestalling further damage to the cherry trees. How well the beavers will survive in another location is highly problematic.

Timber companies and small timberland owners also suffer substantial damage when valuable trees are inundated by beaver flowages and soon die. Annual losses of timber to beaver damage may even run as high as hundreds of millions of dollars nationally.

The first response of the property owner who's suffering damage is to ask the state wildlife agency to fix the problem. This puts the agency in a difficult position, because the owner usually balks when told that the offending animals must be killed. Instead, the complainant wants the beavers caught in a box trap and relocated. While this may seem kind, it's actually quite the opposite.

In order to understand this seeming contradiction, another facet of beaver biology must be explored. Far from being the peaceful creatures which they appear to be, beavers are highly territorial. If a strange beaver comes to their colony, the inhabitants will either kill it or drive it away. Fierce territorial battles abound where beavers are plentiful, and older beavers soon become heavily battle-scarred from their confrontations with interlopers.

Consider, then, the plight of beavers removed from their colony and set adrift. If the distance is short enough, they will simply return to their former home, thereby continuing the original problem. Otherwise they must either try to enter another beaver colony—a recipe for certain disaster—or try to establish another colony. With virtually every available bit of beaver habitat already taken, the chances that they will succeed are infinitesimal. Further, unless they are transported in the spring, there's insufficient time for them to build a dam and lodge and store food before the onset of winter. In summary, a beaver taken from its colony and left somewhere else has been handed an almost certain death sentence from starvation, from the onset of winter, or, in its weakened condition, at the fangs of predators.

Although the trap-and-transport option may keep the property owner from feeling guilty that he or she has been directly responsible for a beaver's death, it would have been far kinder to simply shoot the beaver in the first place. That's a major reason why most state wildlife agencies now refuse to trap and transport live beavers, even though this stance angers some people. These professionals realize that it's far less cruel to kill the beavers outright than to doom them to a slow, painful death.

In a few well-publicized instances, animal rights activists have tried to save beavers by having them live-trapped, neutered by veterinarians, and returned to their former haunts. This, it turns out, is no kindness either, for it runs afoul of yet another aspect of beaver biology. Beavers, as it happens, have a social structure that's very gender-specific; neutering creates chaos in this complex system and renders the colony ineffective for the task of survival.

Despite these conflicts with humans, the beaver is clearly here to stay. On the whole, this is a very good thing. True, some beavers will have to be removed when they cause excessive damage to the works of people, but the beaver's insatiable appetite for building dams and cutting trees provides a wealth of prime habitat for many other wildlife species. Welcome back, beaver!

2

The Misnamed One: The Muskrat

MYTHS

- The muskrat is a close relative of the common, or Norway, rat.
- It's a scaled-down version of the beaver.
- It's called a muskrat because it's a kind of rat with a musky scent.

OUTSIDE ITS NORMAL AQUATIC SURROUNDINGS, THE MUSKRAT *(ONDATRA ZIBETHICA)* IS OFTEN MISIDENTIFIED AS A RAT, THAT IS, EITHER THE NORWAY OR COMMON RAT *(RATTUS NORVEGICUS)* OR THE BLACK OR ROOF RAT *(RATTUS RATTUS)*. On several occasions I've had someone tell me, with either a shudder or with loathing in the voice, "A great, fat rat crossed the road in front of me. It was *huge,* just disgusting!" Or "I looked out my window, and there was the biggest rat I've ever seen coming up out of the water. It was awful!" After a few inquiries about specifics of the animal in question—size, shape, color, and related characteristics—it became clear that the creature being discussed wasn't the hated and feared Norway rat, but the very distantly related and wholly innocuous muskrat.

Rats have earned their reputation as one of mankind's greatest scourges—destructive disease carriers that have afflicted humans down through the centuries. After all, they were the primary carriers of bubonic plague—the fearsome Black Death of the Middle Ages that decimated Europe—as well as a major source of typhus. Further, their destruction of grain supplies and damage to many other things of value to humans have made them even more feared and despised. Norway and black rats are also immigrants from Europe and represent perhaps our most unfortunate importation of nonnative wildlife.

Beaver *(top)*; muskrat

The muskrat, on the other hand, is a native of North America. Far from being highly destructive and a carrier of disease, it's of great benefit to many forms of wildlife and of considerable value to humans. This interesting mid-sized rodent deserves to be more widely recognized and appreciated.

Even those who recognize the muskrat and don't confuse it with the common rat often tend to think of it as a sort of junior edition of the beaver. In fact, the muskrat is very much its own man, so to speak—related to neither Norway rat nor beaver except by virtue of belonging to the order Rodentia. This order, incidentally, comprising some three thousand species worldwide, is the largest of all mammalian orders.

Despite some similarities to beavers in such things as appearance and habitat, muskrats are far more closely related to voles, those plump little short-tailed rodents that most of us call meadow mice. It's not stretching things much to say that the muskrat is a very large, aquatic vole, and some scientists have actually described it in that fashion.

The origin of the muskrat's name itself is fascinating. We might reasonably deduce that it derives from a combination of the muskrat's long, naked tail and the slightly musky odor produced by its scent glands—but we would be wrong! Rather, the name is the product of a peculiar twist of language called *folk etymology*.

Etymology is the study of word derivations, and folk etymology is the modification of an unfamiliar word by incorrect usage into something with more familiar elements. Although European settlers occasionally adopted—usually in somewhat corrupted form—Native American names for creatures with which they were unfamiliar, they tried to avoid that practice wherever possible. Their avoidance took the path either of naming a North American creature for something at least vaguely similar from Europe, or of using folk etymology to transform the name into something "sensible."

In the present instance, the name of this little marsh-dwelling rodent was originally *musquash* in the Algonquian language. That name made little sense to the colonists, who observed that this creature could give off a musky scent and had a long, naked tail a bit like that of a Norway rat. Put these two observations—musk and rat—together and they sounded quite similar to musquash, yet seemed to make perfect sense. Voilà, Monsieur Muskrat!

As already noted, the muskrat looks much like a small beaver except for its tail, which is nothing like the beaver's except that it's hairless. While the beaver's tail is rounded and relatively short, very wide, and flattened top to bottom, the muskrat's is long, quite slender, and flattened from side to side.

The muskrat is also far smaller than the beaver. An adult varies from a foot and a half to a little over two feet long from nose to tip of tail. This total length is deceptive, however, for the long tail consumes nearly 40 percent of it; as a result, adult muskrats weigh only two to four pounds.

Aside from their fondness for water, muskrats lead very different lives from beavers. For one thing, the muskrat's habitat requirements are far less rigid than those of the beaver. Since they don't construct dams, muskrats have no need of suitable dam sites, or for materials for constructing dams. They need very little in the way of water; anything from a huge lake or river to a drainage ditch or farm pond suits this notably unfussy rodent. In fact, muskrats frequently utilize beaver ponds for their habitat, and the two species seem to coexist quite peaceably.

Muskrats also eat a much wider variety of foods than beavers. Whereas beavers are exclusively vegetarian, muskrats feed extensively on freshwater mussels, frogs, crayfish, and similar aquatic creatures whenever these are available. Now and then they even manage to catch a slow or unwary fish. However, their main diet is plant material; cattails (both the shoots and tubers), water lilies, duckweed, pickerel weed, assorted other pond weeds, bulrushes, sweet flag, and a variety of reeds and sedges are prime muskrat food.

Nor is the muskrat constrained by the need for the inner bark of trees that so often forces the beaver to abandon a colony denuded of surrounding trees and brush. No doubt the muskrat's great adaptability in matters of food and habitat accounts for its nearly ubiquitous presence throughout most of North America, from the subarctic to the Gulf of Mexico.

Often the presence of beavers can be detected by gnawed pieces of wood that have floated far downstream from a colony. The first signs of the muskrat's presence are likely to be a bit more subtle. These frequently take the form of cut pieces of cattails and other aquatic plants, floating about or lodged on the shoreline, and these are easily overlooked except by the careful observer. Far more obvious is the sight of some aquatic greenery mysteriously moving across the surface of the water. In the latter case, closer inspection reveals that it's a sort of wildlife version of Birnam Wood coming to high Dunsinane—a muskrat, almost totally submerged, propelling a bunch of cut vegetation to a preferred feeding site.

These favorite feeding spots vary widely. They may be flat rocks, logs, stumps, matted vegetation, or a composite of trampled mud and reeds. Often a heap of discarded mussel shells or the remains of a number of crayfish announce the location as a choice dining spot for the resident muskrat.

Not only are muskrats more flexible than beavers in regard to food and habitat, but they're also less choosy about living quarters. A beaver is compulsive about building a lodge, and digs a tunnel into a bank only as a last resort in situations where constructing a dam or lodge is totally impractical. Even in a large body of water with an outlet that beavers can't possibly dam, the big rodents will fabricate a lodge near shore, provided the shoreline doesn't drop off too abruptly or isn't excessively rocky. A muskrat, on the other hand, usually lives in a burrow by preference, but is perfectly at ease building a house in marshes where there are no convenient banks steep enough for burrowing.

A muskrat burrow can be quite long—up to fifteen feet—angling up and back from just below the surface of the water. With a larger living chamber at the end, and several escape tunnels, the whole affair can be fairly elaborate. Although muskrats are generally very beneficial, their tunneling proclivities sometimes cause damage to dikes and small earthen dams, making them unwelcome residents in some situations.

A muskrat house is much less strong and elaborate than a beaver lodge. These little domed structures are made in autumn by heaping up a mixture of reeds and mud. The muskrat then burrows up into the mass from below the waterline, excavates a chamber, and digs out additional tunnels for escape routes beneath the water.

Next the muskrat erects several smaller, ancillary structures within a few yards of the main lodge. These little affairs, called *pushups,* enable the muskrat to extend its feeding range when the marsh or pond is icebound. In late fall or winter, after the marsh vegetation has died down, these houses and pushups can be seen dotting large marshes, protruding above the water or ice.

Muskrats even manage to find housing in large swamps that lack banks for burrowing and cattails or similar vegetation for constructing houses. In that event, a muskrat will utilize a hollow log or stump, proving once again how adaptable a creature it is.

Muskrats are an extremely important component of many wetlands. Where they're abundant, they consume an enormous amount of aquatic vegetation, particularly emergent species such as cattails. Without these industrious little harvesters, many wetlands would have almost no open areas, because cattails and similar plants would soon choke out most of the open spaces. Indeed, the percentage of open water in a marsh is often highly dependent on the muskrat population.

This propensity for creating openings in wetlands makes the muskrat extremely valuable to other species. Without open areas, ducks, geese, herons, egrets, and numerous other birds would derive little benefit from marshes, shallow ponds, and similar wet areas. Moreover, areas of open water create places for a variety of submerged aquatic plants to grow, thereby contributing greatly to the wetland's diversity.

Unlike beavers, muskrats seem to have little territorial instinct except in connection with their actual homes. This trait makes excellent evolutionary sense when one thinks about it: if muskrats were as fiercely territorial as beavers, they would be in constant conflict, and it would be very difficult for large numbers of them to coexist in the same marsh.

Muskrats remain active throughout the year, even when winter closes like a giant vise on pond and marsh. Mostly they remain under the ice, foraging for a variety of foods while using their living quarters and any pushups they may have built as feeding spots. Several muskrats frequently share the same winter quarters, particularly in their domed houses of mud and reeds. Evidently this communal arrangement serves two important purposes: first, it keeps the interior of the house relatively warm; and, second, the warmth helps to keep the water from freezing in the interior entrances to their underwater exits.

Although muskrats normally don't attempt forays above the ice, they may sally forth if they run short of edibles, as sometimes happens in small wetland areas. Then they'll leave the water and travel overland, even in the dead of winter, to seek another food source. Also, the need to find a mate may cause lone muskrats to leave their winter quarters in late February or March and set out on a cross-country journey. At such times they may turn up in odd places, such as garages, or may be killed while crossing highways.

Heidi, our black Labrador retriever, and I once had a painful adventure with one of these winter wanderers. I happened to glance idly out the window one afternoon when the snow lay deep on the field adjoining the house. A distant movement caught my eye, and I spied a brown object moving in our direction. As it drew closer, the object soon resolved itself into a muskrat.

Intrigued by such an incongruous sight, I took Heidi and went outdoors for a closer look. The muskrat continued to move steadily in our direction until it was perhaps twenty feet away. Then, without warning, it accelerated and charged straight at us at top speed.

Astounded by this turn of events, I failed to react in time. Heidi, curious about this strange creature, lowered her head, and the onrushing muskrat bit the unfortunate dog savagely on the end of the nose! Blood, which always looks far worse on snow, seemed to fly everywhere, while the muskrat turned and fled in the direction it had come from. Because I had to attend to Heidi's wounded nose, we never did learn what became of the muskrat.

What was a muskrat doing at that time of year, exposing itself to danger in open, snow-covered fields? There was no rabies in Vermont at the time, so any suspicion of a rabid animal can be eliminated. No doubt the peripatetic muskrat had either been forced to seek a new food supply or was heeding the siren call of the mating season. At any rate, it provided a thoroughly memorable experience, albeit a most unpleasant one for the unhappy Heidi!

As befits their semiaquatic status, muskrats have a full complement of adaptations to equip them for life in the water. Like beavers, muskrats are notable divers without benefit of outsized lungs. With the same sort of adaptations possessed by beavers, they can easily spend ten minutes underwater, and longer dives of up to fifteen minutes are by no means unknown. One biologist saw a muskrat dive and remain submerged for an astounding seventeen minutes, come to the surface for just three seconds, and then dive for another ten minutes!

The muskrat has other useful adaptations as well. Its hind feet are partially webbed to provide efficient paddles for easy movement in the water, while its long tail acts as a rudder. On occasion, this tail can present a rather comical appearance: when the muskrat is floating at rest, it sometimes angles its tail upward, completely out of water. There, unsupported, it forms a shallow curve, first upward and then, farther back, gradually drooping toward the surface.

Meanwhile, the muskrat's admirable fur coat protects its owner against the effects of even the most frigid water. Although muskrat fur isn't quite as highly prized as that of the beaver, it's nonetheless very handsome. Beneath the long, dark brown, glossy outer guard hairs lies a dense coat of fine, soft, grayish underfur, designed to keep water away from the muskrat's skin.

Because of the high quality of their pelts, muskrats have long been a staple of the fur trade in North America. In Wisconsin, for example, the value of muskrat pelts from 1970 to 1981 exceeded $33 million. Despite being both extensively and intensively trapped each year, however, the muskrat has remained abundant down through the years and continues to thrive. This is due to a combination of the muskrat's adaptability and its exceptional reproductive capacity, which is far more like that of the vole than the beaver.

For an animal of its size—two to four pounds—the muskrat has a remarkably short gestation period, lasting a little less than a month. Further, it has at least two litters annually, often three, and sometimes as many as four, with most litters consisting of four to eight young. Add the fact that a female muskrat can breed as early as eight months, and the biological potential of a pair of muskrats, if completely unchecked, is astronomical!

Obviously, many influences restrain the growth of muskrat populations: otherwise we'd be knee-deep in muskrats. Disease, parasites, injuries, and predators all take a toll. Humans are now a major predator of muskrats, but mink are also an important natural enemy. Thoroughly at home in the water, mink often enter muskrat houses in search of a meal. There, despite being much smaller than an adult muskrat, these fierce little predators quickly do away with the occupants and dine in style. Otters, too, even swifter and more at home in the water than mink, sometimes prey on muskrats.

Other predators also abound, though they mostly prey on young muskrats. Big snapping turtles and large fish, such as northern pike and garfish, relish young muskrats and feed on them at every opportunity. Raccoons, hawks, great horned owls, and alligators also esteem muskrats as food. All in all, this system of muskrat fecundity and numerous predators seems to work very well, as muskrats are neither scarce nor overabundant in most locations.

If there's any threat to the muskrat, it's habitat loss. Despite its adaptability, the little marsh dweller can't survive without suitable areas for food and shelter. Habitat loss comes from two sources. The first is the continued drainage of wetlands for everything from agricultural land to development. Although laws protecting wetlands have helped to slow the destruction of our enormously important wetlands, they haven't yet halted it.

The second major source of habitat loss for muskrats comes from the nutria, or coypu *(Myocastor coypus),* a rather large rodent that weighs fifteen to twenty pounds. It was unwisely imported from South America for its fur, and its introduction has had serious repercussions for both marsh and marsh dwellers. The nutria is found mostly in coastal areas in Maryland and Delaware, from Georgia to Louisiana and Texas, and in parts of California, Washington, and Oregon. Inland, there are also some infested areas in parts of the Midwest.

The problems with this interloper are twofold: first, the muskrat can't compete with the nutria for food; and, second, the nutria's feeding habits simply destroy a marsh. Where muskrats eat the stems and leaves of aquatic plants, as well as some tubers and fleshy roots, the nutria totally destroys the root systems of these plants. Once the vegetation is eliminated, there's nothing to hold the land in place, and the marsh is washed away.

Louisiana, where coastal wetlands have enormous economic value not only for muskrats, but also for alligators, shrimp, and waterfowl, is losing

300,000 acres of these priceless wetlands annually to nutria "eat-outs." The state is fighting back, however: in addition to promoting the use of nutria fur in an effort to control these destructive pests, Louisiana is also trying to develop a market for nutria meat. And in Delaware, the state has recently received a $2-million federal grant to eradicate the nutria, if possible.

Despite the habitat destruction wrought by humans and nutria, an enormous amount of suitable habitat remains in North America for the versatile muskrat. Considering its enormous reproductive potential, and the vast acreage of available habitat, it seems likely that this fur-bearing rodent will continue to be a common sight throughout most of North America.

Red squirrel *(top);* gray squirrel

3

A Sylvan Odd Couple: The Red Squirrel and the Gray

MYTHS

- Red squirrels drive out gray squirrels.
- Red squirrels castrate gray squirrels.
- Gray squirrels remember where they bury nuts.
- Squirrels are vegetarians.
- Gray squirrels are very tame.

THE RED SQUIRREL *(TAMIASCIURUS HUDSONICUS)* AND THE EASTERN GRAY SQUIRREL *(SCIURUS CAROLINENSIS)* ARE BOTH TREE SQUIRRELS AND LIVE IN PROXIMITY TO EACH OTHER THROUGHOUT A WIDE STRETCH OF TERRITORY, YET THEY DIFFER WIDELY IN SIZE, MODE OF LIVING, FOOD AND HABITAT REQUIREMENTS, AND MOST ESPECIALLY IN PERSONALITY.

These two tree-dependent rodents inhabit the same tracts of forest land from the southern edge of the eastern Canadian provinces down the Atlantic coast to Virginia, west to Illinois, and back north through parts of the Dakotas; narrow bands also extend southward into Kentucky, Tennessee, and the Carolinas. All or parts of some five Canadian provinces and twenty-four of the lower forty-eight states harbor both red and gray squirrels.

Within this broad area, two of the most common and enduring of all wildlife myths have sprung up. It's an article of faith with many people that "the reds drive out the grays," and, even worse, that the reds castrate the male grays.

While it's true that red squirrels sometimes chase their big relatives, they hardly do it with the intention of "driving them out" in the sense of excluding them from a large area. Far more than gray squirrels, reds are highly territorial in regard to their dens and caches of food. Thus, if a gray squirrel approaches one of these too closely, the red may chase it for a short distance. Once the gray vacates the small area around den or cache, however, the red has no further interest in pursuing it.

This behavior is similar to what is often seen in birds. It's a common and entertaining sight to see small songbirds chasing and harassing much bigger birds, such as crows, that have violated the airspace above the songbirds' nests; this behavior even extends to attacks on hawks and other birds of prey that fly too close to a nest site. In both cases—squirrels and birds—the aggrieved party seems to have the upper hand, and the offending party flees. In the same manner, gray squirrels will drive away reds that come too close to their dens.

Many people say they've seen red squirrels chase grays away from backyard bird and squirrel feeders. No doubt they have, in some cases, but this behavior is far from universal. In years of watching the interaction of the two species at our feeders, only twice have we seen a red squirrel take after a gray.

In one instance the gray ran up a tree, pursued by the red. When the red got too close, the gray turned on it, and the red squirrel fled down the tree. A moment later the red made another try; this time it took only a menacing move of the gray squirrel's head to make the red hurriedly depart for good. In another incident, we watched a red run toward a gray eating fallen seeds beneath a feeder. The gray ran a short distance, although not in much haste; the red pursued for three or four feet and then turned back to feed, clearly uninterested in chasing the gray any farther.

In contrast to these two incidents, we've observed countless occasions when grays have pushed reds out of our feeders or chased them away from a choice spot under a feeder. In general, we've seen the reds defer to the much bigger grays in numerous ways that make it abundantly clear that reds don't "drive out the grays."

It's also likely that if a hungry red squirrel approaches a gray squirrel satiated from stuffing itself with sunflower seeds, the gray may allow itself to be chased out of a feeder. At that point the red squirrel is powerfully motivated by hunger, while the gray has no incentive to remain or to act aggressive.

All of these observations and suppositions aside, there is irrefutable proof that red squirrels don't drive out the grays. If they did, as one biologist pointed out, there would be few if any gray squirrels left in the very large areas where the ranges of the two species overlap. Obviously this has not happened. Both species are common throughout most of these areas, and go their separate ways with only occasional minor conflicts.

Because red squirrels do chase grays on occasion, the erroneous conclusion that the reds drive out the grays is at least partially understandable. However, the myth that the reds castrate the male grays must have originated either as the result of an overheated imagination or as a deliberate tall tale.

Merely consider the facts. The gray squirrel generally weighs from two to three times as much as the little red. Even what are normally the most peaceable of animals will fight savagely, if necessary, to protect themselves. Nor could a red squirrel, with its little teeth, neatly snip off the testicles of the gray with one or two bites. The notion that the much bigger gray squirrel would allow its testicles to be gnawed off by its little relative is preposterous; long before that happened, the gray would make squirrel hash of the offending red!

The final proof, if any is needed, is the same as for the myth that red squirrels drive out the grays. If the reds were successfully sterilizing the grays, there would soon be few grays within the red squirrel's range—and that clearly isn't the case.

Grays are not only much longer and heavier than the little reds, but their size is exaggerated by their long, bushy tail. This plumelike affair can be quite splendid on a healthy specimen, partially because the gray's tail is longer in proportion to head and body than the red's tail. When a gray is sitting up, its beautiful tail follows the line of its back as far as the head and then curves backward in a graceful arc. This characteristic elicited Winifred Welles's perceptive imagery in *Silver for Midas:*

> *My squirrel with his tail curved up*
> *Like half a silver lyre.*

Both species shed their coats twice a year and simultaneously grow new ones, first in late spring and again in early autumn. As in the case of birds, this process is known as molting. Oddly enough, the tail in both species molts only once, in the middle of summer.

Considering how often these two squirrels can be found in the same sections of woodland, it's worthwhile—and quite fascinating—to see how different their lives are in many respects. Food requirements and feeding habits are as good a place as any to start.

In simplest terms, gray squirrels depend on nuts as their most essential food supply; historically, wherever nuts—acorns, hickories, beechnuts, butternuts, and black walnuts—are abundant, so are gray squirrels. Red squirrels, on the other hand, rely heavily on the seeds and buds of coniferous trees: pine, spruce, fir, and hemlock. Thus, even though they relish nuts where available, their range is confined mostly to coniferous or mixed-growth forests. In a nutshell (pun intended), that is the most important difference in the food requirements of the two species.

Of course, nature is never quite that simple, so there are many complexities and permutations of the simple formula that nuts equal gray squirrels, while cones equal reds. Red squirrels eagerly devour nuts when they're available and can subsist on them in the absence of seeds from cones. The converse isn't true, however; gray squirrels don't feed on the seeds from cones, and so are dependent on nuts. In addition, a wide range of foods is shared by the two species. Fungi, fruit, many kinds of seeds, hardwood buds, sap, and the inner bark of trees all are part of the diets of both reds and grays.

Both species also share another food source, one that isn't commonly thought of in connection with squirrels. Far from being vegetarians, as most people believe, reds and grays alike are quick to consume birds' eggs and baby birds whenever they have the chance, and grays are known to feed on forest-dwelling frogs. The reds, even more omnivorous than the grays, go a step further and often eat the young of small mammals such as mice and voles.

Just as the basic food requirements for the two species are widely divergent, so too are their methods of caching food and consuming it. Gray squirrels are justly noted for industriously storing nuts underground. These are buried singly in dispersed fashion throughout each squirrel's territory.

This behavior has led to the common myth that gray squirrels remember where they've buried each nut. Research has shown that grays don't have that kind of remarkable memory; although they may recall general areas where they've buried nuts, the location of individual nuts isn't part of their memory bank.

That being the case, how do squirrels find the nuts when they need food? The answer lies in their remarkable sense of smell. Gray squirrels locate buried nuts—even under many inches of snow—by scent alone! This means that individual squirrels dig up nuts buried by other squirrels wherever they happen to smell them. Such behavior, though it might seem a bit unjust, probably results in something like a fair exchange in most cases.

This uncertain and somewhat haphazard retrieval of buried nuts has one vital, if unintended, result. Many nuts are never found, and these sprout and grow into new nut trees. While some nuts, such as the acorns of white oaks, will sprout on the surface of the ground, others must be buried in order to germinate.

Red oak acorns germinate much better underground, and burial is almost a necessity for butternuts, black walnuts, and hickories. Indeed, it's estimated that perhaps 95 percent of all hickory trees come from nuts unwittingly planted by squirrels. Thus the relationship between the gray squirrel and nut trees is really symbiotic: the squirrels require nuts in order to survive, and the nut trees require squirrels to plant them. As an added complexity, researchers have learned that gray squirrels promptly consume white oak acorns, which sprout soon after they fall, but bury red oak acorns, which need to spend the winter underground in order to sprout effectively in the spring. Again, this arrangement benefits both tree and squirrel. What a superb example of nature's intricate interrelationships!

Red squirrels also cache food, but in a far different manner. Their strategy is to fill storehouses—a hollow tree or log, or an underground area—with as much food as they can pack away for the long winter months. Particularly where the seeds of coniferous trees are their main food supply, these caches can be enormous, sometimes numbering two or three thousand cones. These represent such a fine supply of seeds that the U.S. Forest Service sometimes appropriates them for use in starting conifer seedlings. This is not as harsh as it sounds, for the stolen cones are replaced with corn or other suitable food to see the little hoarder through the winter.

One of the more amusing sights in nature is to watch a red squirrel transporting pine or fir cones to its cache, or from cache to feeding site. Usually the squirrel grasps the cone by one end with its teeth, so that the little creature looks for all the world as if it's puffing on a grotesquely large cigar!

No picture of squirrel food sources would be complete without mention of backyard feeders. The popularity of feeders has exploded in the past few years, particularly in suburban areas, though a great many rural residents have feeders, as well. The proliferation of rich supplies of sunflower seeds and other avian goodies has had a major and largely unintended effect in helping gray squirrels expand both their numbers and range in areas where nuts are scarce or almost nonexistent.

Squirrels—both red and gray—at feeders are extremely controversial. Humans who provide and stock feeders divide neatly into two camps: those who loathe squirrels and those who enjoy them. My wife and I are among the latter.

Those who detest squirrels at their feeders will go to almost any lengths to defeat them. Businesses are built on the manufacture and sale of "squirrel-proof" feeders (some more successful than others), and at least one book has been written on the subject. On the other side, there is an organization for squirrel lovers.

Without question, squirrels consume an awful lot of birdseed, creating an added expense that the homeowner may not want. It's also undeniable that squirrels are astoundingly clever at circumventing obstacles designed to keep them out of feeders. It's a little like trying to pick a lock: it can be done, but not easily! Thus, some people regard squirrels as unwanted, expensive nuisances and take umbrage when the squirrels defeat their anti-squirrel efforts.

On the other hand, many people enjoy the antics of squirrels as much as they appreciate the beauty and variety of the birds attracted to their feeders. To them, the extra seed consumed is a small price to pay for the double pleasure of observing both birds and squirrels. Ultimately it's a highly personal decision, and there's room in the world for both squirrel haters and squirrel lovers.

The actual eating habits of the two species are as different as their principal food sources. Gray squirrels eat nuts pretty much where they find them—here, there, and everywhere. Reds, on the other hand, have favorite feeding places, frequently on stumps or a fallen log. There, huge middens of cone scales build up, as revealing of the squirrel's presence and habits as are the kitchen middens of human origin, so eagerly sought by archaeologists. Such squirrel middens will commonly fill a bushel basket or more.

Watching a red squirrel shuck a cone to get at the seeds is an edifying experience nearly as humorous as watching it transport the cone. Holding the cone in its front paws, the squirrel gnaws with astonishing rapidity, cone scales flying in every direction. Simultaneously, the front paws rotate the cone in a manner highly reminiscent of a human eating corn on the cob. In an amazingly short time, the cone has been reduced to flat scales on the ground, while the seeds now reside inside the squirrel.

Red squirrels are also great hands for temporarily storing certain kinds of food in trees. When a late-summer mushroom or an apple, some distance from an apple tree, is seen securely placed in a low crotch, the culprit is almost always an industrious little red.

Allusion has already been made to the very different personalities of these two species, but nowhere is this more evident than in their vocalizations. In keeping with their generally more sedate deportment, gray squirrels speak sparingly. Aside from high-pitched alarm calls of young squirrels being attacked by a predator, the gray's voice is confined to something that is usually—though inadequately—described as a bark. Although these sounds are of short duration, they are nothing like the bark of a dog. Rather, they're thin and raspy, somewhat akin to the sound of a rough file being drawn quickly over a surface.

Red squirrels are an entirely different proposition. Far more hyperactive than their big relatives, the little reds have a wide range of vocalizations—

most of them seemingly expressive of particularly vile and opprobrious thoughts!

The characteristic red squirrel sound most often heard at a distance is a long, high-pitched *chirrrrrrrr* that can last for more than ten seconds. This sound can be approximated by trilling an *r* continuously on a high note. Far more interesting, however, are the red's other sounds, which can best be appreciated at short range as part of a full-scale audiovisual presentation.

Red squirrels love to scold. At the slightest opportunity they indiscriminately scold humans, birds, and animals that are unfortunate enough to impinge on what the reds regard as their exclusive domain—but most of all they revel in scolding other red squirrels.

A red squirrel, seething with indignation, has a veritable arsenal of squirrelish invective to hurl at an intruder. In fact, these little squirrels have raised indignation to the level of high art! Chirks, squeaks, grunts, and other sounds too difficult to describe pour forth in incredible profusion from that tiny body, accompanied all the while by jerks of the tail and shuffling and stamping of the hind feet. Some of these sounds—particularly a very high squeak and a low, raspy grunt—seem to be made simultaneously, although they may simply follow each other so rapidly as to give that illusion.

A red in the throes of one of these vituperative displays brings to mind a tiny teakettle as it boils over, hopping on the stove while bubbling and steaming furiously, or a miniature Vesuvius about to blow its top. Heard and seen at close range, this is a performance to be treasured! Because of this feisty behavior, incongruous in a creature so small, we affectionately refer to any red squirrel around our home as Big Red.

The personality difference between gray and red squirrels shows in another major way, too: wariness. To those who have encountered gray squirrels only around backyard feeders or as furry mendicants begging for food in urban parks, it may come as a surprise that completely wild gray squirrels are extremely wary.

When a gray squirrel sights a human in the wild, it either dashes into a handy den or conceals itself in the top of a tree. There it's almost impossible to see, even after the autumn leaves have fallen. Even if the person then sits down and remains absolutely still, the squirrel usually won't reappear for about twenty minutes. In rural areas, gray squirrels are widely hunted for their delicious meat, and any squirrel hunter can testify to the wariness of grays in the wild.

Red squirrels in the wild, on the other hand, are at least as apt to remain and roundly curse a human intruder as they are to hide. Even with the semi-tame squirrels around backyard feeders, some of the same behavioral differences are evident. For example, we have only to open our door and the gray squirrels flee precipitously, streaking for the nearby trees. The feisty little reds, however, will stay, sometimes even when approached within two or three feet, and hurl insults at us like an angry fishwife whose toes have just been stepped on.

Another amusing trait is the red squirrels' penchant for closely pursuing each other. A common sight in the woods is to see one red squirrel racing after another at top speed across the ground, then chasing it around and around a tree trunk in a sharp upward spiral. At some point the whole affair is reversed, and the two spiral down the tree and speed away out of sight. All the while, the two squirrels will maintain the same distance of a foot or two from each other, so perfectly synchronized that their capers resemble the operation of some mechanized toy.

Living accommodations for the two species differ greatly. Though both like tree dens, the grays are also great summer nest builders. The presence of gray squirrels is often revealed by these great, rounded bunches of leaves and twigs, sometimes several to a tree. Such nests, called *dreys,* are usually placed away from the trunk, where the juncture of two or more branches makes a convenient platform. Dreys are constructed so as to shed water. They're hollow, with an entrance hole, and, like nests in tree cavities, are lined with whatever soft materials are available.

Grays tend to utilize several tree cavities—sometimes as many as seven—if they're available, but usually they have both dens and dreys. Once the young have been born, the mother will frequently move them from one den or nest to another.

This may be partly a defensive strategy, making it more difficult for predators to dine on baby squirrel. Biologists also speculate, however, that it may be a way of avoiding the major flea infestations—sometimes thousands in a den—that regularly plague squirrels. Indeed, one squirrel biologist who sported a full beard was finally forced to shave it off because so many squirrel fleas hopped into it!

Red squirrels also den in tree cavities; when these are lacking, they'll build dreys, though in much smaller numbers than the grays. However, the reds are just as apt to den in a burrow in the ground, often dug between the roots of a tree or under their food cache. When they do build nests, these are placed close to the trunk of a tree and are much smaller and less conspicuous than the nests of the grays. Usually the red's nest is constructed of twigs, or twigs mixed with shredded bark, with far fewer leaves than the gray's leafy bower.

Red squirrels remain highly active most of the time during the winter, though they may sometimes hole up for a day or two in exceptionally stormy or bitter weather. They are always solitary in the winter, and may den in a tree cavity, in a burrow, or simply in tunnels in the snow.

Reds also like to expropriate birdhouses for a den. While cleaning out our bluebird houses one spring, I failed to pay sufficient attention to the fact that the house was full of various soft plant materials. As I dug down into the mass, I was startled to feel movement and hear an angry squabbling sound. At the first glimpse of red fur, I realized that I had wrecked the happy winter home of a red squirrel, which was now decidedly indignant!

Gray squirrels also remain active throughout the winter, although they may be semi-dormant for a couple of days during unusually inclement weather. Unlike the solitary reds, grays also den communally. These dens are shared by males and juveniles, which are very sociable and may groom each other. Breeding females, on the other hand, are cantankerous and den by themselves. Gray squirrels in the more northern parts of their range normally have one litter a year. Farther south, two litters a year is common. Breeding for the first litter is in January and February, with May and June the usual breeding time if there's to be a second litter. The young are born blind and naked, usually two or three to a litter, although there may occasionally be as many as five.

Red squirrels have somewhat larger litters, usually four or five, but sometimes as many as seven. They, too, sometimes bear two litters a year, although one is more common.

If humans handle young squirrels in their den, the mother's exceptionally keen sense of smell detects it. However, rather than abandoning her brood, she immediately moves her young to another den as a precautionary measure.

As this behavior indicates, squirrels of both species are extremely good mothers. Despite the care that they lavish on their broods, however, life is apt to be brief and hard for squirrels in the wild—juveniles and adults alike. Eighty percent of young squirrels die during their first year, and adults seldom live more than three or four years.

One major reason for such a high mortality rate is predation, for many hungry creatures are perfectly happy to dine on squirrel. Foxes, coyotes, and bobcats sometimes catch an unwary squirrel on the ground, but members of the weasel family are a much greater threat to tree squirrels (see chapter 14). Although weasels themselves spend most of their time on the ground, they can climb trees to get into squirrel dens, and these slender little predators can easily slip into any hole or burrow that a red squirrel can enter. Two larger weasel cousins, the marten and the fisher—especially the former—can also pursue and capture squirrels in the treetops.

Hawks and owls also take a heavy toll of squirrels, and I received a striking demonstration of that fact within a few feet of our house. I had just opened the door and stepped out, when a movement caught my eye. No more than fifteen feet away, a hawk—either a small Cooper's hawk or a large sharp-shinned—had just seized Big Red (actually one of our many Big Reds) in its talons and was bearing it away toward the nearby woods. And although barred owls, at least, seem generally to prefer mice to squirrels (see chapter 11), a fellow naturalist told me of watching a barred owl in the forest seize a red squirrel.

Predation is by no means the only danger faced by squirrels. A variety of parasites and diseases can also prove fatal, and accidents happen, as well. Remarkably, however, in view of lives spent racing around through the trees, leaping from slender branch to slender branch, squirrels are rarely injured in falls. Partly this is because squirrels are so agile and adept in the trees that they seldom fall, and partly it's due to their ability to survive a fall without serious injury.

My mother had a bird feeder at her upper-story bedroom window, and she frequently became incensed at the sight of hog-fat gray squirrels gobbling up the seeds she intended for the birds. At such times she would shoo the grays off the feeder. They, without the slightest hesitation, would launch themselves outward to fall two and a half stories onto the bare ground below. Not once did one of these daredevils ever show any visible sign of injury. No doubt this ability is due to a combination of light weight and substantial wind resistance, which, much in the manner of a parachute, slows the squirrel's descent greatly. This wind resistance, in turn, is the result of the squirrel's long, bushy tail and the way it flattens itself out, much in the manner of a flying squirrel.

Despite the dangers imposed by all these hazards, both gray squirrels and reds are prolific enough to cope with this steady attrition and assure their continued abundance. It seems likely, therefore, that this tree-dwelling odd couple will continue to frequent our forests, just as they've been doing successfully for so many thousands of years.

4

Nature's Gliding Machines: Flying Squirrels

MYTHS

- Flying squirrels truly fly.
- Flying squirrels are active only at night.

ALTHOUGH THEY'RE COMMON ENOUGH IN MANY FORESTED AREAS, THESE LITTLE CREATURES ARE SEEN SO SELDOM THAT THEY MIGHT ALMOST BE CONSIDERED WOODLAND GHOSTS. At night, perhaps, car headlights may for a fleeting second reveal what looks like a large, square leaf passing overhead, leaving the occupants wondering exactly what it was that they saw. More than likely, they glimpsed a flying squirrel—possibly the only view they would ever have of one, outside a museum or zoo. Reclusive as these little rodents are, however, it's very much worth the effort to learn about them, for they are wonderfully adapted for the sort of life that they lead.

The most basic fact about flying squirrels is that they don't! That is, despite their name, flying squirrels don't truly fly, as bats and birds do. Rather, they use a remarkable adaptation that enables them to glide, and they could more accurately be called *gliding* squirrels.

And just how can a squirrel manage to glide? Very easily and well, as it turns out, thanks to a most unusual and extraordinarily useful adaptation. A wide, loose flap of fur-covered skin extends outward from each side of the flying squirrel's body and stretches from the front ankle back to the rear one. When the squirrel wishes to "fly," it simply launches itself into the air from high up in a tree and spreads its legs to the widest extent possible. This action stretches out the flap of skin on each side, making the squirrel resemble a large, square pancake with head, feet, and a tail.

Northern flying squirrel

Stretched out in this fashion, the squirrel, which weighs very little, has a large surface area to support it. This adaptation enables the creature to glide laterally for a considerable distance before reaching the ground or a lower point on another tree.

Just how far can a flying squirrel glide in terms of horizontal distance? There's no absolute answer to this question because there are so many variables. These include the height from which the squirrel launched; the slope of the land, if any; and wind velocity and direction. Height is the most important of these, because the glide ratio of a flying squirrel is about three feet horizontally for each foot of height.

Our North American flying squirrels can certainly glide two hundred feet or more, and some authorities say as much as three hundred feet. Considering the glide ratio, that would mean launching from a height of one hundred feet—certainly possible in many situations, although most glides are a good deal shorter. In theory, at least, the only limit to a flying squirrel's length of glide is the height from which it started.

Because a flying squirrel in a long glide acquires a certain amount of velocity, one might think that the little animal would injure itself by crashing into its landing site on the trunk of a tree. But in fact this presents no problem for the squirrel, which, as it prepares to land, drops its tail and raises its forepaws, thus creating wind resistance much as an airplane does when it lowers its flaps during a landing. The squirrel, now almost in a vertical position, lands lightly against the target tree and promptly scurries off about its business.

Nor is a soft landing the only talent that flying squirrels exhibit while gliding. By using its tail as its main rudder, and also by moving its legs, thereby tightening or loosening the flaps of skin on one side or the other, the flying squirrel can maneuver well enough to avoid branches and other obstacles during a glide. In fact, a flying squirrel has been observed banking through an arc of as much as 180 degrees during a glide.

It's widely assumed that flying squirrels are almost entirely nocturnal, but twice I've seen the little creatures abroad in daylight. One afternoon, while it was still full daylight, I saw a flying squirrel glide from a large maple tree to a tall balsam fir on the edge of our woods. The squirrel landed a few feet above the ground, scampered up the trunk, and, to my surprise, disappeared into a small hole that I hadn't previously noticed. Though I watched for some time, the squirrel failed to reappear; almost certainly the hole was the entrance to its den. Where it had come from or what it had been doing is anyone's guess.

On another bright, sunny day, two of us were doing forestry work. As we sat on the ground that noon, munching our sandwiches, my companion suddenly whispered, "Look, there's a flying squirrel coming toward us."

Sure enough, a flying squirrel on the ground was headed in our direction; its movements, impeded somewhat by the skin flaps, were far less graceful than the glides normally associated with its species. There was a small clearing in front of us, with mushrooms growing here and there, and these fungi were an evident attraction for the little squirrel. While we watched, it eagerly ate parts of several mushrooms, gradually drawing closer to us.

Paul Fiske, my partner, had a pair of leather work gloves which he had laid on the ground beside him. Slowly and quietly slipping them on, he whispered, "I'm going to try to catch it."

Moving with great caution, Paul got to his feet; then he leaped up and ran at full speed toward the squirrel. With nothing close by to climb, the squirrel could only run for the nearest tree. Hampered by its skin flaps, the squirrel was a slow runner, and Paul scooped it up after a few long strides. I ran to join him, and together we inspected the soft, big-eyed creature that Paul was holding very gently. After we had satisfied our curiosity, the squirrel was placed on the ground and released, whereupon it ran to the nearest tree, climbed it, and glided off into the forest.

Unquestionably, flying squirrels are predominantly nocturnal. Because I've seen them abroad twice in full daylight, however, it's evident that these furry gliders are more diurnal than is generally believed. After all, my two daylight encounters with them can hardly be unique, and no doubt many other people have occasionally observed flying squirrels active in the daytime.

There are two species of flying squirrels native to North America, the northern flying squirrel *(Glaucomys sabrinus)* and the southern flying squirrel *(Glaucomys volans),* as well as some subspecies. Although these two species are somewhat different in size and in some of their habits, they're very similar in at least three characteristics.

First, the eyes are large, very dark, and prominent—hence the genus name of *Glaucomys.* This name ultimately stems from the Greek word *glaukos,* meaning bluish gray, probably because of the way the flying squirrel's eyes reflect light. Second, both species have wonderfully fine, soft fur, far softer than the fur of their tree squirrel relatives, such as the red and the gray. And third, of course, both species glide whenever possible, and are much more at home in the trees than on the ground.

Despite these major similarities, there are a few noteworthy differences between the two species. Size is one. The northern flying squirrel can have a total length of as much as thirteen inches, although eleven inches or so would be more typical. Of this, the tail takes up roughly four to five inches, while the remainder is head and body. In total length, then, the northern flying squirrel is only about an inch shorter than the red squirrel, although it appears considerably smaller. That's because the red is stockier and weighs from 24 to 50 percent more; the red weighs in at seven to eight ounces or more, while the northern flying squirrel only tips the scales at a distinctly lightweight four to six ounces.

The little southern flying squirrel is considerably smaller than its more northerly cousin. In fact, it's the smallest of all our North American tree squirrels. With a total length of about eight to nine inches—three to four and a half inches of it being tail—this diminutive glider weighs only two to two and a half ounces. By way of comparison, this is about half of the northern flying squirrel's weight, and only one-third of the red squirrel's.

Even our northern flying squirrel is small, however, when compared to some of the giant Asian flying squirrels. One, the red giant flying squirrel *(Petaurista petaurista)* of Southeast Asia, has a combined head and body length of about sixteen inches, plus a tail slightly longer than that—a total length approaching three feet! So large is this squirrel that its thick pelt of mahogany-red hair is occasionally a commercial item. This squirrel can reputedly glide for very long distances, but *Walker's Mammals of the World,* a highly respected source, only lists a known glide distance of about 225 feet.

Although the ranges of our two native flying squirrels show considerable overlap, their habits and habitat requirements vary considerably. The northern flying squirrel, as its name implies, inhabits much of Canada and Alaska. It also spills down into the United States through New England and New York and all the way along the spine of the Appalachians into Tennessee and a bit of western North Carolina; into the northern portions of the Great Lakes states; southward through the Rocky Mountains; and down the Pacific coast into a considerable piece of California.

In contrast, the southern flying squirrel is found in much of New England and nearly all the rest of the eastern United States as far west as the edge of the Great Plains. These ranges alone are a clue to the different habitat and food requirements of the two species. In the far north, where much of the northern flying squirrel's habitat lies, forests are almost exclusively coniferous. Conversely, throughout much of the southern flying squirrel's range, deciduous forests are very much the rule. Where their territories overlap, forests contain areas of conifers, mixed growth, and deciduous stands.

These differences in habitat have considerable implications as far as food is concerned. In much of the northern flying squirrel's range, nut trees are absent. Although this squirrel devours nuts avidly when they're available, it can make do without them. In summer, lichens and fungi make up much of its diet; in winter, the northern flying squirrel is known to raid the caches of red squirrels. In the latter instance, it seems doubtful that the red even realizes it's being robbed, since it is sound asleep when the nocturnal robber pillages its hoard of cones. Arboreal lichens are also an important winter food source.

Berries, seeds, and fruit are also prominent in the northern flying squirrel's diet. Although we don't think of flying squirrels as carnivorous, like red and gray squirrels (see chapter 3) they eat meat whenever it's available. Mostly, their meat supply consists of nestling birds (they consume the eggs, as well) and the young of mice, voles, and shrews.

Not unexpectedly, southern flying squirrels are more dependent on nuts, but they eat many of the same foods—fungi, seeds, fruit, and berries. They also consume a substantial amount of meat. In fact, the gentle appearance of this tiny glider is quite deceptive, and a number of experts regard it as the most carnivorous of all the tree squirrels. Nestlings and birds' eggs, carrion, baby mice and voles, and even adult mice and shrews are killed and consumed by this little terror. Moreover, as it turns out, the southern flying squirrel is considerably more aggressive than its much larger northern cousin and generally is dominant wherever the two species inhabit the same tract of forest.

There are other differences between the two species, as well. For example, the northern flying squirrel molts twice a year, shedding its fur in May or June, and donning a heavier winter coat in September. The tail, however, only molts once, in early summer. The southern flying squirrel, on the other hand, molts only once, in September.

There are also differences in reproduction and nesting habits. Northern flying squirrels usually have only one litter a year—an eminently sensible system, considering the all-too-brief summer throughout much of this squirrel's range. They breed in late winter and, after a forty-day gestation period, give birth to between two and five blind, naked young.

In contrast, southern flying squirrels usually have two litters a year throughout much of their range—one in the period from February to May, the second from July to September. The dual litters reflect the much longer warm season available to the southern flying squirrel, which takes full reproductive advantage of this climatic benefit. After a forty-day gestation, they give birth to tiny, naked, blind young, each weighing less than one-quarter ounce.

As in the majority of mammals, male flying squirrels of both species do nothing to help raise their offspring. The females, however, are excellent mothers that defend their young and are known to move them to a new nest if parasites become too abundant in the original nest.

Living quarters and winter habits for the two species also vary. In summer, the northern flying squirrel builds summer nests, usually close to the trunk of an evergreen tree. In winter, however, they reside in tree cavities, often enlarged from the prior labors of woodpeckers.

In far northern climes, these squirrels often hollow out growths known as witches' brooms. These peculiar masses are caused by a fungus that sometimes infects spruce and balsam fir trees. The fungus causes the tree to grow a dense, tangled maze of tiny branches that somewhat resembles an old-fashioned broom—hence the name. This hollow in a witch's broom, after being heavily lined with grass, feathers, or other soft material by the enterprising occupant, is evidently warmer than a tree cavity lined with soft material.

Flying squirrels of both species are quite sociable in winter, and as many as eight northern flying squirrels of the same sex may share winter quarters, providing warmth for each other in the den. In extremely cold weather, northern flying squirrels semihibernate, often sleeping through several days until the weather moderates.

Southern flying squirrels also build summer nests, although these are leaf nests in hardwood trees such as oaks and hickories. Although they remain active all winter, and eschew the occasional semihibernation of their cousins, southern flying squirrels are even more social than their northern counterparts when it comes to communal winter quarters. Twenty or more have been known to occupy the same winter home, and one observer reported that *fifty* southern flying squirrels shared a single tree cavity in Illinois!

Although flying squirrels generally make little noise, they can vocalize, though the sounds are a bit different for each species. The northern flying squirrel emits low chirps and sometimes makes little clucking sounds when upset. The southern species, on the other hand, twitters, chirps, and utters high-pitched sounds sometimes described as "tseets."

As might be expected of small, largely defenseless creatures, flying squirrels of both species are a target for a wide variety of predators. Owls of various sorts are a major predator; according to one Alaska biologist, a pair of nesting northern spotted owls can cause the demise of as many as 440 flying squirrels in a single year.

Hawks are also listed as a predator of both species, but if flying squirrels are indeed totally nocturnal, as some sources indicate, it seems inconsistent for those same sources to list the completely diurnal hawks as their predator. However, this inconsistency aside, hawks probably do catch an occasional flying squirrel from those that now and then forage in daylight hours.

Far more serious predators than hawks abound, however. Weasels, martens, fishers, foxes, coyotes, raccoons, snakes, and bobcats all take their toll. That leaves one more important predator, at least where humans live—the house cat.

I can personally testify to the efficacy of these small felines when it comes to catching flying squirrels. Until it unfortunately burned, there was a very large barn on our farm. Built in 1866, it was forty-five feet wide, seventy feet long, three stories high, and was full of hollow partitions and a multitude of nooks and crannies. An amazing assortment of wildlife inhabited that barn, including flying squirrels.

Although we never saw a live flying squirrel in the barn, we often found evidence of their presence. Our house cats regularly prowled the area, sometimes during daylight hours and sometimes at night. Often when we entered the barn we would see the tails of squirrels that the cats had killed and eaten. The majority of them were from red squirrels, but it was by no means uncommon to find the soft, silky tails of flying squirrels—mute testimony both to the abundance of the squirrels and the predatory prowess of the cats.

It's perhaps appropriate to mention that two subspecies of the northern flying squirrel are considered endangered, according to the U.S. Fish and Wildlife Service. One is the Carolina northern flying squirrel *(Glaucomys sabrinus coloratus),* which is found in just five locations—three in western North Carolina and two in eastern Tennessee. Its closest relative is the other endangered subspecies, the Virginia northern flying squirrel *(Glaucomys sabrinus fuscus);* this latter is found in just a few areas of Virginia and West Virginia.

While we hope that these two endangered subspecies survive, we can also take comfort from the fact that both the northern and southern flying squirrels are generally doing quite well throughout the major part of their respective ranges. Despite the attrition from disease and a horde of predators, the little gliders are sufficiently elusive, and have a high enough reproductive rate, to ensure their survival.

5

A "Pig," Perhaps, but Not a Hog: The Porcupine

MYTHS

- Porcupine quills are barbed.
- Porcupine quills are filled with air, like a balloon.
- Porcupines can throw their quills.
- A porcupine is a hedgehog.
- Porcupines are adept at climbing trees.

THE PORCUPINE IS A CURIOSITY, AN ODDITY, ONE OF THOSE ABERRATIONS DECIDEDLY OUT OF THE MAMMALIAN MAINSTREAM. Slow, plodding, awkward, rather dim-witted, and quite lacking in adaptability, the porcupine would seem to qualify as a poor candidate for survival in a dangerous world where so many seemingly better-equipped animals have passed into extinction. Yet despite such numerous disadvantages, this walking collection of anomalies has survived for millions of years and gives every appearance of wending its bumbling, unconcerned way into the distant future.

The saving grace in this oddball animal's makeup is, of course, its quilly armor. Without that wonderful evolutionary quirk, the porcupine would long since have gone the way of dinosaurs, woolly mammoths, and sundry other unfortunates. However, the porky's defenses are so formidable that, with few exceptions, it can waddle its way through life with minimal problems from would-be predators.

Porcupine quills are actually highly specialized hairs, totally distinct from the short underfur and very long guard hairs. A porcupine has an astonishing 30,000 quills to protect its back, flanks, and tail—an average of about 140 quills per square inch on much of its body! The face and underside lack

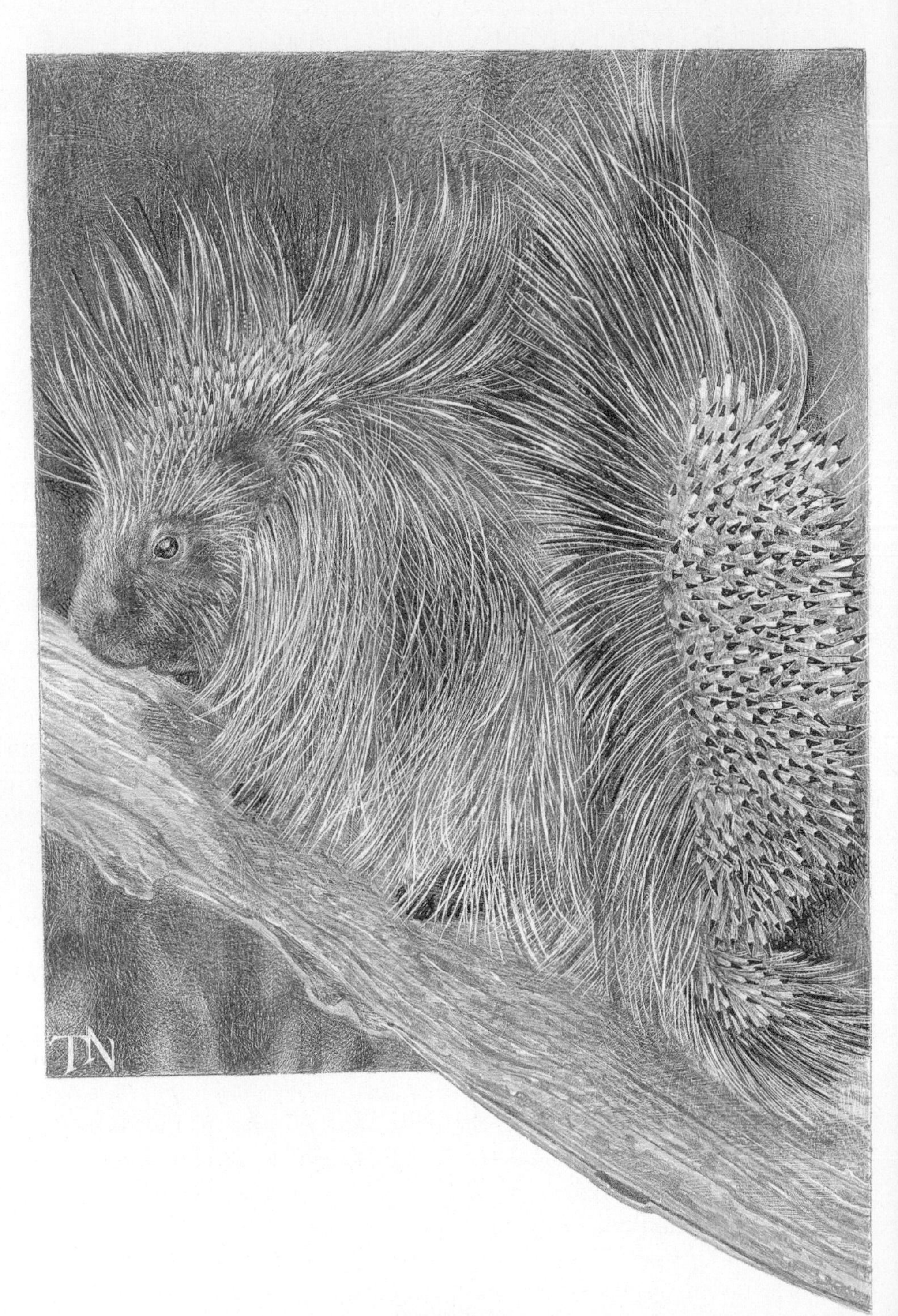

Porcupine

quills, however, thereby giving porkies the animal equivalent of an Achilles' heel, albeit one that few predators are able to take advantage of.

The quills vary in length; those on the back are the longest, while the shorter ones are located on the head and tail. The longest quills can span as much as five inches, but most measure about three inches. The majority of quills are white for much of their length—possibly to make the porcupine's weapons highly visible and hence more threatening to a prospective predator—but a few are completely black. Each quill has an extremely sharp tip and is very loosely attached at its base; it hardly takes more than a touch to embed the sharp point in the flesh of an attacker, whereupon the quill easily departs from its very tenuous attachment to its owner.

It's widely believed that porcupine quills are equipped with tiny barbs, like those on a fishhook: they aren't, although for all practical purposes they might as well be. Instead, the tip of a quill has diminutive scales, much like tiny fish scales. These overlap so that the raised edges point toward the rear and function in much the same manner as the barbs on a fishhook; once embedded in flesh, these scales make it very difficult and painful to remove a quill. In any event, to the animal or human painfully stabbed by a number of these miniature lances, the technical difference between barbs and scales is probably of no consequence!

Because of the scales, lodged quills tend to work their way forward. There are numerous instances, in both humans and animals, of quills that have disappeared, worked forward, and, after a period of time, emerged at a considerable distance from the entry point. Occasionally, quills have been known to work their way into a vital organ and cause death. This is probably relatively rare, however; a greater danger is that an animal whose face and mouth are riddled with quills may be unable to hunt or eat properly, and thus may die of malnutrition.

One myth about porcupine quills stems from the fact that their hard, tough exterior conceals a more or less hollow interior. The operative term here is "more or less." Folk wisdom to the contrary, quills aren't filled with air like a balloon, and hence don't deflate when their posterior end is cut off. Instead, quills are filled with light, spongy material, and cutting the end of the quill off is no aid to removing it from a dog or other creature. Indeed, this tactic may even make matters worse, since it leaves less quill to grip with pliers or forceps, and may splinter the quill as well.

Undoubtedly, the oldest, most enduring, and most widely believed myth about porcupines is that they can throw their quills. Just when and how this fiction began is buried in the mists of time. However, it most certainly was given great credence by that famous Roman, Pliny the Elder, who died in the cataclysmic eruption of Vesuvius that buried Pompeii and Herculaneum in A.D. 79.

Pliny's massive, thirty-seven-volume *Historiae Naturalis XXXVII,* published in A.D. 77, is riddled with scientific inaccuracies—men whose feet were turned the wrong way, winged horses, unicorns, and many other nonexistent marvels. Evidently, Pliny was an uncritical believer in the wide variety of fanciful tales that reached his ears, and one of these concerned the porcupine. In 1601, Englishman Philemon Holland, also a true believer, gave us this translation: "The porkpen hath the longer sharp pointed quilles, and those, when he stretcheth his skin, he sendeth and shooteth from him."

Perhaps the porcupine's formidable tail has something to do with this myth. When an intruder approaches a porcupine too closely, the animal reacts by lashing its tail ferociously from side to side. One quick flick from that spiky caudal appendage, and the unwary can find themselves with a horde of quills driven deep into their soft flesh. While this clearly isn't a case of throwing quills, it might seem that way to a victim of the tail's rapid swipes.

Although porcupines lack the ability to throw their quills, they do have the capacity to raise and lower them at will. When threatened, a porky elevates its quills so that it bristles like some sort of mammalian cactus with giant spines. Shakespeare was evidently aware of this trait. In *Hamlet,* the ghost of Hamlet's father says, "I could a tale unfold whose lightest word . . . would freeze thy young blood . . . make thy knotted and combined locks to part, and each particular hair to stand on end, like quills upon the fretful porpentine."

Although not quite as quaint as Holland's "porkpen," this is yet another interesting and—to us moderns, at least—amusing variation of the porcupine's name. Regardless of old-fashioned spellings, the derivation of this walking pincushion's name is as intriguing as the creature itself: it descends from Italian *porcospino,* meaning "spiny pig" (*porco* = pig + *spino* = thorn or spine). Evidently some Italian or another thought that the porcupine either resembled a pig in some fashion or was related to it. In any event, the *spino* was properly descriptive, even though the *porco* missed its mark by a wide margin, and the name stuck.

Far less appropriate is the term *hedgehog,* by which the porcupine is widely known. A hog it is not, nor does it frequent hedges. Wholly unrelated, except by virtue of being mammals, the porcupine and hedgehog bear little resemblance other than the highly superficial characteristic of a spiny exterior.

For openers, the porcupine is a rodent, whereas the hedgehog is an insectivore. Then there is the matter of size: while porcupines are rather hefty animals, the hedgehog is diminutive—only about ten inches long. Even in the matter of their defensive weaponry, the two species are more divergent than might appear at first glance. The spines on the hedgehog, although sharp enough to give it good protection when it curls into a tight ball if danger threatens, are hardly the lethal weapons of the porcupine's detachable quills.

Why, then, did the North American porcupine come to be widely known by the misnomer "hedgehog"? The first English colonists in North America were quite likely unfamiliar with porcupines except from literary references, since Old World porcupines are absent from Great Britain and most of Europe. These colonists had a penchant for calling unfamiliar creatures by the names of familiar ones from "back home," and in a prickly sort of way the porcupine bears at least a superficial resemblance to the hedgehog, a common denizen of English hedgerows.

Regardless of names, accurate or otherwise, the porcupine's arsenal of quills is so formidable that most predators quickly learn the desirability of giving the porky a wide berth. A variety of predators may occasionally kill a porcupine: these include bobcats, bears, cougars, wolves, coyotes, and wolverines. However, after one or two painful encounters, these predators emerge sadder and a great deal wiser in the ways of porcupines. The single exception seems to be the domestic dog; while some dogs learn to shun porcupines after one or two episodes, others never seem to learn.

That leaves only two predators—humans and fishers—to control the porcupine population. In the absence of fishers for a time in many parts of the porcupine's range (see chapter 14), humans attempted to control the burgeoning numbers of what had become a destructive nuisance. Mostly these efforts were failures.

A favorite tactic was to pay a bounty for dead porcupines. As with any bounty, the theory was that the money would induce people to go forth and expend great effort to bring about the demise of large numbers of porkies. And as with other bounties, the porcupine bounty proved a failure, mostly because—as any professional wildlife manager will affirm—bounties simply pay people for killing animals that they would kill anyway.

A rule of thumb regarding bounties is that in order to be effective, they have to be set so high as to be prohibitively expensive. Generally, porcupine bounties were established at somewhere between twenty-five and seventy-five cents—rarely, perhaps a dollar—per animal. Although this was a good deal more money back in the 1940s and 1950s than it is now, it was hardly munificent enough to generate any major effort to kill porcupines.

There was one intriguing twist to the bounty business, however. Evidence that a porcupine had been killed was required before payment was made, and the animal's two ears were the usual proof specified in the bounty law. There used to be persistent rumors that certain enterprising individuals, possessed of somewhat elastic consciences, would cut earlike shapes from the unquilled portions of the porkies. When wizened by a suitable period of drying, the "ears" would be presented as the genuine article to the town clerk or other agent authorized to make payment. These individuals, rarely expert in such arcane fields as porcupine anatomy, were unlikely to question the evidence. Thus, supposedly, an unscrupulous few multiplied their bounty money severalfold. Whether this practice actually occurred, at least on any substantial scale, or is merely the result of rural rumor and folklore, is now buried in the past.

At any rate, the return of fishers to their former habitat proved vastly more effective, and much less expensive, than bounties or other human efforts to control excess numbers of porcupines. Partially as a result of smaller porcupine populations, and partly as a result of the disfavor into which bounties in general have quite properly descended, porcupine bounties have mostly—and perhaps entirely—been abandoned.

As already mentioned, porcupines are rodents—relatives of mice, squirrels, rabbits, and woodchucks. Moreover, the porky is a large rodent, second in size only to the beaver among its North American relatives. The normal weight for an adult porcupine ranges from fifteen to twenty pounds, but on rare occasions a big male can reach as much as twenty-eight to thirty pounds. At that size, a porky up a tree can, at a first hasty glance, be mistaken for a bear cub, although a second look will quickly correct that error.

Like all other rodents, porcupines are noted for their gnawing—a trait that often gets them in trouble, especially during times of overpopulation. All rodents have four incisors, two in the upper jaw and two in the lower. A porcupine's incisors are large and sharp, just right for cutting through tree bark to get at the tender inner layer, or for demolishing anything that contains traces of salt; their incisors also possess the interesting quality of being orange in color.

A porcupine's normal diet is apt to be deficient in salt, a lack that causes it to seek out anything that can remedy the deficiency. Certain types of plants, natural salt licks, and—more recently—the residue from winter salting of roads to melt snow and ice are as eagerly consumed by porcupines as fresh fruit and vegetables once were by sailors suffering from scurvy. But although consumption of road salt may benefit the porcupine's health, the interaction between porcupines and automobiles that often results definitely does not—and many are the porkies that end up as roadkill because of it!

Although the inner bark of trees may be the porcupine's dietary staple in winter, these big rodents consume a surprising variety of foods during the warmer months. Grasses, clover, buds, succulent water plants, apples and other fruit, and a number of herbaceous plants, acorns, and raspberry shoots, among others, are all grist for the porky's mill.

In winter, the porcupine's diet becomes far more limited. As already noted, the inner bark of trees is a mainstay, but hemlock twigs are also a favored winter food throughout that tree's range. Indeed, the presence of a porcupine can often be detected in winter by the large number of hemlock twigs and small branches littering the snow beneath a prime feeding site. Buds of various hardwood species also form an important component of the winter diet.

Favorite feed trees are usually easy to identify, especially if several porcupines live nearby. Branches are apt to be stripped of foliage and small twigs, and large areas of bark are frequently removed. This latter phenomenon can result in girdling and killing the tree, and thus a major overpopulation of porcupines can seriously damage a forest.

The porcupine's fondness for salt, which amounts almost to an obsession, has already been noted. This characteristic has done little to endear the prickly beasts to humans. Anything that contains traces of salt from sweat is fair game for the porky: ax and other tool handles, canoe paddles, sweaty clothing, gloves, and numerous other items are all eagerly consumed by the sodium-hungry creatures.

The list also includes a few rather bizarre items that might not readily come to mind, and thereby hangs an amusing tale—although it didn't seem so amusing at the time. Our 1854 farmhouse had a privy (no longer in use) attached to the house in an enclosed corner of the side porch. At the time we bought the house, in the early 1960s, Vermont's overpopulation of porcupines was at its height, and encounters with porkies were frequent.

One night my wife and I were awakened from deep slumber by the sounds of great and terrible munching, as if an entire army of rats and mice had descended on us and were consuming our dwelling. On investigation, we found that a porcupine had somehow gotten in under the base of the outhouse, climbed the ten feet or so to the top, and was contentedly chewing away, getting its ration of salt from the seats of the two-holer!

Nor is that the end of the list of strange things that porcupines gnaw for their salt. People hiking the Appalachian Trail from Vermont's Lincoln Mountain have sometimes returned to find their automobiles disabled because porcupines gnawed through various hoses coated with winter road salt. Indeed, signs there have even warned motorists about this potential hazard.

Few things in the natural world seem to elicit as much mildly ribald speculation and comment as the mating habits of the porcupine: How do they "do it"? Frequently the question is rhetorical, followed by the rather hackneyed answer, "Carefully, very carefully."

Actually, it's only humans, not porcupines, who make a big deal of all this. When mating time arrives in early autumn, males begin to act uncharacteristically aggressive toward each other, threatening movements and gestures are made, and occasional fights may even ensue. Once a male has established dominance, he may mate with more than one female.

When the actual moment of mating arrives, imagination far outstrips reality; contrary to much rumor and folklore, the whole affair is remarkably simple. The female porky stands on all fours, flattens her quills (remember that the quills are movable), raises her hindquarters, and moves her tail to one side. Presented with such an inviting opportunity, the male mounts her from the rear without the slightest discomfort. Voilà! Mission accomplished.

An extraordinarily long gestation period, at least for a mammal of the porcupine's rather modest size, follows—a full seven months. By comparison, a coyote's gestation is only about two months, and that of the beaver just over four months.

In yet another display of their highly individual qualities, porcupines bear only a single young each year. However, the little porky weighs a solid pound at birth, which is a lot of baby for a younger, smaller female that may only weigh twelve pounds or so.

This lone baby, born in April or May, is known by the thoroughly delightful name of *porcupette*. Its eyes are open at birth, and it has a full complement of tiny quills that it seems to know instinctively how to use. The quills present no difficulties during birth, since they're very soft, but they dry quickly, and the porcupette rapidly becomes armed and dangerous.

From unhappy personal experience, I can vouch for the validity of that statement. Years ago, when I was doing forestry work, my black Labrador, Heidi, often accompanied me in the woods. I never had to keep track of her, for she never wandered far from me, so I let her go about her business while I went about mine.

On that particular spring day, we were far back in the forest, I marking timber and Heidi reveling in the newly bare ground with all its wonderful fresh scents. Suddenly she ran up to me, and I looked down to speak to her. To my horror, I saw that she had carefully retrieved a very live and healthy porcupette!

At my command, Heidi obediently dropped the little creature, which could hardly have been more than two or three days old, and I was able to survey the very considerable damage. Her nose, muzzle, and the inside of her mouth were all riddled with tiny quills—dozens and dozens of them. There was only one thing to do: Heidi and I immediately departed from the forest and made a hasty journey to the vet's, where he tranquilized the unfortunate dog in order to remove the quills as painlessly as possible.

About two years later my wife felt something odd while she was patting Heidi's neck. Upon investigation, she found an emerging quill—clearly one that had traveled back from her mouth. This is yet another example of a quill's propensity for working its way through flesh.

A porcupette is quite precocious. In addition to being born furred, armed with quills, and with eyes open, it can climb and eat solid food only hours after birth. Like other baby mammals, however, the porcupette's main source of food is its mother's milk. The mother often leaves her baby tucked safely away in a den or snug nook, going off by herself to forage while the porcupette stays where it was placed.

The young porky is weaned after a couple of months, but mother and baby remain together for a few more weeks. Then the young one wanders off to lead its mostly solitary sylvan life, where it will reach sexual maturity at age three.

Among the many porcupine oddities is the fact that females are far more territorial than males and vigorously defend their turf against other females. Further, it's the young females, rather than the young males, that disperse widely in the fall. In both instances, this is the opposite of the behavior displayed by most mammals.

Despite the severity of the climate throughout much of their range, porcupines aren't hibernators and remain active all winter. They do, however, seek shelter in a den of some sort—usually a small cave, a hollow tree, or a protected cranny far back in a tumble of rocks. They may remain in the den for a day or two during exceptionally inclement weather, but emerge to feed as soon as the weather improves a little.

Although porcupines are generally solitary animals for most of the year, they seem gregarious enough during the winter denning period. Where they're sufficiently numerous, several at a time may share a den, evidently coexisting quite peaceably until spring.

With their bulky bodies and short legs, porkies are ill-suited to plowing through deep, soft snow, so their winter feeding forays are apt to cover only very short distances. This means that trees near a den—especially a communal one—may take a beating and eventually die. While this isn't a serious problem when porkies are in balance with their habitat, overpopulation can cause substantial damage for some distance around major den sites.

Although porcupines aren't usually particularly vocal, they have a considerable repertoire of sounds they can call on when the occasion demands it. Several years ago, for instance, I came across a porky track in the snow. Curious to know where it led, I followed it for a short distance to a huge old sugar maple, a good three feet in diameter, which had broken off at ground level and blown down. In the middle of the trunk, between the broken-off roots, was a large hollow leading for some distance up into the trunk. There the porky had made its den.

The tracks clearly indicated that the porky was at home, so I peered up into the hollow, where, in the dim interior light, I could just make out the occupant's tail and broad rump, completely filling the cavity. Like Queen Victoria, the occupant was not amused.

The porcupine was well aware of my presence, and no doubt had detected my approach long before I reached the den; perhaps to compensate for conspicuously bad eyesight, porcupines have excellent hearing and a keen sense of smell. Displeased, the porky began to emit a whole range of sounds—squeaks, grunts, groans, snuffles, and assorted other odd noises. As it continued to mutter, mumble, and squeak to itself, I was reminded of nothing quite so much as a querulous old man, suddenly awakened from his nap, who peevishly protests the intrusion. The performance was so ridiculous that I actually burst out laughing!

Porcupines can be vocal at other times as well. Robert Brander, a National Park Service ecologist who has done extensive research on porcupines, has documented a fascinating phenomenon. In his research, Brander found that as many as a dozen porcupines may congregate in mid-to-late summer at a chosen location. Usually this is an old clearing that is growing up and provides a good source of food.

Brander believes that socialization, rather than food, may be the primary purpose of these gatherings. Coming together in this fashion, he postulates, may be a useful prelude to the breeding season, which follows in roughly another month. In any event, Brander reports that the assembled porkies are extremely vocal, uttering a wide variety of sounds.

In addition, porcupines are also reported to have a very shrill, high-pitched cry that some have described as a scream. Possibly this is one of the sources of the "screams" often erroneously attributed to bobcats.

Over the years, I've had many memorable and often amusing encounters with porcupines. However, what was certainly the most bizarre occurred within a year or two of our dog's contretemps with the porcupette. It happened in this fashion.

I was driving past a dairy farm and, having grown up on such a farm, slowed down to gain a better view of a particularly nice-looking herd of Holsteins milling around a feed bunker. To my dismay, a porcupine suddenly appeared, waddling unconcernedly toward the cows, which were some distance away and still hadn't noticed their novel visitor.

I had no usable weapon in my car, and, fearful of the result if cows and porcupine came together, I wheeled into the farm's dooryard and knocked on the farmhouse door. The young farm wife appeared, and I hastily introduced myself and explained the impending disaster. She was quite properly suspicious of me at first, but I finally convinced her that I was, in fact, a forester employed by the state.

Unfortunately, her husband and the hired help had gone off on some errand. I inquired if there was a gun handy. She quickly found one, but didn't know where her husband kept the ammunition. I then told her that, in a pinch, a stout club would do, and she finally located a baseball bat. Grabbing the bat, I ran for the pasture.

Alas, all of this had taken time—too much time. When I reached the pasture, it was evident at a glance that my worst fears had been realized. The cows were clustered around an extremely dead porcupine, trampled by the hooves of very large bovines, while the fleshy, bulbous noses of several of the Holsteins were stuck full of innumerable quills!

I returned the baseball bat to the wife and broke the bad news to her. I've often wondered since then whether the farmer endeavored to pull the quills himself or had to call a veterinarian to tranquilize the beasts before extracting the porky's darts. Knowing the tremendous strength of an upset cow, I'd bet on the latter.

Porcupines are full of surprises. One of the most recent concerns their skill in tree climbing. Porkies have always been regarded as slow, awkward, but nonetheless very able climbers, since they spend much of their lives in trees. Lately, however, biologists have learned that it's not uncommon for porcupines to fall out of trees and injure or even kill themselves!

Possibly some of the porky's arboreal difficulties stem from its somewhat unusual method of locomotion. Whereas most mammals grip a tree with their front feet and propel themselves with their hind legs, porcupines do exactly the opposite. As it holds the tree with five strong, sharp claws on each hind foot, a porky reaches up, one front foot at a time, sinks the four front claws into the bark, and hauls itself upward in hand-over-hand fashion. Descent reverses this mode, as the porcupine climbs down tailfirst.

Now that the fisher has brought porcupine numbers down to a normal level, encounters with the clumsy rodents have become uncommon enough to be distinctly pleasurable under most circumstances. My most recent experience with a porky is a good example.

While deer hunting in a few inches of fresh snow, I spotted suspicious-looking tracks some distance away. I hastened to investigate, and even before reaching the tracks, my nose told me that it belonged to a porcupine. Porkies tend to have a rather pungent odor, much of which stems from living in a den carpeted with porcupine dung, and the slight breeze wafted the telltale scent to my nostrils from a distance of several feet.

The opportunity was irresistible, and I followed the broad trail, complete with tail-drag marks, for two or three hundred yards. There the trail ended at a hemlock of modest proportions. Gazing up into the tree, I searched and searched for the track maker, which, unless it had sprouted wings, was surely in the tree.

Finally, after looking futilely for some time, I happened to glance at a spot far down the tree, where I was startled to see the porky on a branch barely above my reach. This was a wonderful chance to observe a porcupine at close range, so I stood there and watched.

The porky seemed totally unfazed by my presence. It was nipping off the tips of hemlock twigs and masticating them in slow but purposeful fashion, and it seemed to have no intention of interrupting that useful activity. From time to time it paused for a few seconds to gaze incuriously at me with its dull shoe-button eyes—then resumed its munching.

For quite some time I stood there, talking softly to the porky, which of course paid not the slightest attention to my meaningless sounds. Finally, with considerable reluctance, I went on my way, leaving the porcupine to continue its contented feeding. The experience brightened my day greatly, though it's doubtful that it made any great impression on the porky.

Although porcupines have often been regarded as intolerable nuisances, it hasn't always been thus. For example, at one time—either by law or custom—it was considered bad form to kill a porcupine in areas of extensive, trackless forests. The theory behind this was that the porcupine was the only animal that a lost and unarmed person could easily kill for food in order to survive. In fact, a porcupine *is* easy to kill: a sharp blow on the end of the nose with a club or stout stick will dispatch one with ease. It's unknown how many lost souls were actually saved from starvation in this fashion—probably very few—but the theory at least sounded good!

Native Americans had an even better opinion of the porcupine, and held it in high esteem. In some of their legends, the porcupine is something of a hero, and indeed, they found the porcupine to be a most useful creature. It was easy to kill, and its meat was deemed a delicacy, an opinion confirmed by a number of people in recent times. Further, its quills, often dyed, were widely used to make handsome decorative designs on such things as baskets and canoes, as well as a form of jewelry for personal adornment.

Our own perception of the porcupine is changing, too, especially now that porkies are no longer present in excessive numbers. True, a dog may occasionally run afoul of the quill-bearing rodent, or a porky may cause damage by gnawing on anything from a tree to a tool handle. Nonetheless, more and more we're recognizing the plodding, seemingly imperturbable porcupine as an important component of the forest ecosystem; more significant, we're perhaps beginning to fully appreciate the unique qualities of this mammalian curiosity.

Big brown bat

6

Evasion Beats Entanglement: Bats

MYTHS

- Bats are blind.
- Bats fly into people's hair.
- Bats pose a major threat of rabies to humans.
- Bats are a sort of a flying mouse.

MONSTROUS BEYOND IMAGINING, ALL-CONSUMING, BLACKER THAN BLACKEST NIGHT, THE HIDEOUS SATAN IN THE *NIGHT ON BALD MOUNTAIN* SECTION OF WALT DISNEY'S *FANTASIA* SPREADS GIGANTIC BAT WINGS AS IT TURNS FIERY EYES TOWARD THE LOST SOULS ABOUT TO BE ENGULFED IN WRATH AND FLAMES. This batlike depiction should come as no surprise; it's merely another manifestation of the fear, horror, and superstition with which bats have been regarded down through the ages.

In any drawing of a haunted house, bats are likely to be seen emanating from its towers and windows. Bats were also regarded as "familiars"—that is, spirit helpers in animal form—of witches. Evil witches in conical hats—toothless hags on airborne broomsticks—usually are shown with a flight of companion bats, like a swarm of night fighter planes shepherding a heavy bomber on its deadly journey. Indeed, one unfortunate woman in fourteenth-century France was burned as a witch for no better reason than the abundance of bats around her home!

Bats have also figured prominently in potions and curses. "Wool of bat" is a key ingredient in the witches' brew in Shakespeare's *Macbeth*—a concoction that surely rates as the most unappetizing cookery of all time! And in

The Tempest, Caliban includes bats with such things as beetles and toads when invoking his curse on Prospero.

Even in this modern age of supposed enlightenment, millions of people still shudder at the vision of bats flying at their heads to entangle themselves in human hair, or of bats as fiendish vampires sucking the blood out of their victims. To all of these has been added the overblown fear of rabid bats.

All of this is truly sad, both for bats and humans. Because of superstition and irrational fears, bats are among the most persecuted of all our mammals. We humans, too, are impoverished by these archaic attitudes, for bats are unquestionably among the most astonishing and fascinating of all living things.

So complex and intriguing are bats that it's difficult to know where to begin in telling their story. However, the origins of bats and the structure that gives them the ability to fly represent as good a starting point as any.

Approximately 55 million years ago, and nearly that long before we humans even existed, bats had evolved into winged creatures—the only mammal that truly flies. The earliest known fossils are from the early Eocene period. In geological terms, this is a mere ten million years or so after the dinosaurs died out.

These early bats, such as *Archeonycteris, Icaronycteris, Hassianycteris,* and *Paleochiropteryx,* were very similar to their modern counterparts—smallish bats about the size of our present-day North American bats. This doesn't mean, of course, that bats didn't evolve from flightless ancestors; rather, it simply means that we haven't yet—and may never—find their intermediate ancestors. Most bats are, after all, very small, fragile creatures with delicate bones, which doesn't make them the best of subjects for fossil preservation. Also, most of the very early bats were in tropical regions, where decay takes place with astonishing rapidity and there's not much of the sedimentation needed to preserve fossils.

There is evidence of evolutionary processes at work in these earliest bat fossils, however. *Icaronycteris* and *Archeonycteris* appear to be somewhat more primitive in ear and throat structures than *Hassianycteris* and *Paleochiropteryx.*

Whatever the evolutionary path, bats developed a unique wing structure. Whereas birds and pterosaurs (or pterodactyls, if you wish) evolved with wings supported by the bones of the arm and a single finger, a bat's wing is supported by the arm and *four* greatly elongated fingers. Thus the order to which bats—and only bats—belong is Chiroptera, Greek for "hand-wing."

The bat's fifth finger, or "thumb," incidentally, far from being elongated in the manner of its other digits, is a small hook used for climbing or walking. Yes, even though flight is their normal mode of locomotion, bats actually do climb such things as cave walls and trees, and even walk along the ground on occasion when seeking food. Neither are bats helpless in the water, for competent observers have even observed them swimming.

Flexible, yet strong for their extremely light weight, the elongated finger bones form a perfect support system for the bat's wings. The actual wings consist of a wonderful double membrane of skin, similar to very thin, pliable leather. These wings not only cover the elongated finger bones, but also continue rearward until they actually attach to the tiny hind legs. Another section of this membrane connects one hind foot to the other; in most species, this rear section of membrane extends back to the tip of the tail. In so-called freetail bats, however, the slender tail itself projects well beyond the tail membrane.

This design of a bat's wing can be compared to that of an airplane. Just as an airplane wing has a thin skin of aluminum on the top and bottom, covering a network of ribs and struts to lend rigidity, bats have a skin covering over the top and bottom of their wing bones, with blood vessels and nerves in the space between.

The result of all this is that bats have a very large airfoil in comparison to their weight. In terms of human technology, this makes them somewhat analogous to the old-fashioned biplanes. A biplane's very large ratio of wing surface to weight made it slow but highly maneuverable. In the same manner, bats are fairly slow fliers compared with many birds, but they have incredible maneuverability.

Have you ever watched—*really* watched—bats in flight? Few people have; most who view bats from time to time either fear and loathe them or at best give them only cursory attention. This is unfortunate, because bats are truly amazing fliers! My wife and I often sit on our patio at dusk to watch the "bat show." The land slopes away steeply below our patio, so that bats at eye level and above are silhouetted against the evening sky. Thus we're able to enjoy the well-nigh incredible maneuvers which these little fliers make in pursuit of their prey.

This leads us to the method that most of our native bats use while hunting. Although the expression "blind as a bat" is a common one, it's at least as inaccurate as it is common. Different species of bat vary somewhat in the quality of their vision, but no bats are blind, and most actually see quite well.

However, when flying in the dark while searching for tiny insects, even excellent night vision has serious limitations. Instead, bats rely on a most remarkable system, similar to radar or sonar, known as *echolocation*. To describe this system in its most basic terms, a bat in flight emits high-frequency sound pulses, above the range of human hearing. When these pulses strike nearby objects, their "echoes" return to the bat, which, with its incredible hearing, identifies and locates these objects with uncanny precision. This "radar" system is so amazingly sophisticated that scientists still lack anything approaching a full understanding of it.

The eighteenth-century Italian biologist Lazzaro Spallanzani was the first to postulate that bats could navigate in total darkness through a mysterious sense connected to their hearing. His views were generally ridiculed until the 1930s, however, when Donald R. Griffin, a Harvard University bat biologist and researcher, confirmed Spallanzani's beliefs and identified the mysterious sense as echolocation.

However, even though scientists are still learning about echolocation, they do know that it depends on two components, one physical and the other biological. In the physical component, a bat emits high-pitched sound waves. When these waves strike an object, they "echo"—that is, bounce back to be received by the bat's incredibly sensitive ears. Then the biological component of echolocation takes over, as the bat translates this information with astonishing accuracy and rapidity into an aural picture of its surroundings.

Not only is each nearby object placed precisely in spatial terms, but its size and form are also evident to the bat. Thus a bat can instantly determine whether its system has locked on a human, a branch, a tiny mosquito, or a larger, slow-moving moth. Moreover, it is constantly assimilating this stream of information not for one object, but for all of the many objects around it. So remarkably rapid and precise is this echolocation system that a bat can fly through a maze of wires strung throughout a totally dark room!

Even this description, however, does a grave injustice to the astonishing complexity of a bat's echolocation system. Imagine a jet fighter plane—a marvel of human ingenuity and technology—patrolling the skies through the impenetrable gloom of night. Constantly searching for a possible enemy, its radar sends forth radio waves at a relatively low frequency that return somewhat limited information but can "see" objects at long range.

Suddenly this lower-frequency radar strikes what might be an approaching enemy. As the distance between the planes decreases, the pilot shifts to radio waves of much higher frequency. Although these lack the long range of the earlier signal, they provide a much more detailed picture of the potential enemy.

An amazing human achievement? Substitute sound waves for radio waves, and insects for an enemy plane, and bats have been doing this for countless millions of years! In fact, bats that hunt by echolocation use *three* different systems, depending on the species of bat.

One system uses *constant frequency,* usually referred to as CF. A CF bat, as such bats are known, utilizes brief, intermittent sound bursts at a given frequency. A typical frequency for a CF bat might be 115 kilohertz. This, incidentally, is far beyond the range of human hearing, which extends only to about 20 kHz. So-called CF bats are found in Europe, Asia, and Australia.

Our North American bats use different systems, however. Some, like the southwestern myotis *(Myotis auriculos)* use frequency modulation and are known as FM bats. Bats using this method of echolocation emit bursts of sound that sweep through a wide range of frequencies in an astonishingly brief period. A typical signal from an FM bat might sweep from 100 kHz to 50 kHz in just two *thousandths* of a second! The FM signals that these bats use are the sound-wave equivalent of radio FM signals.

Other North American bats use a mix of CF and FM sounds, very much in the manner described in the jet fighter analogy. These bats are called CF-FM bats. When the longer-range CF sounds of a CF-FM bat, such as the western red bat of Arizona *(Lasiurus blossevillii),* detect potential prey, the bat zeroes in on its target with FM sweeps that give it a much more detailed idea of the prey's texture.

Because most bat sounds are far above our range of hearing, we fail to realize the extraordinary intensity of their calls. For example, a common species such as the big brown bat *(Eptesicus fuscus)* generates some 110 decibels of sound at a distance of four inches. This is as loud as the alarm on a smoke detector at the same distance; we just can't hear it! Animals with higher-range hearing than ours, such as domestic dogs and their wild kin, can undoubtedly hear some of the lower-pitched sounds emitted by bats; whether they can hear bat sounds in the upper range of 100 kHz and beyond is highly questionable, however.

To further add to the amazing variety and complexity of bats' echolocation, consider that some bats which snatch their prey from the ground or from foliage use *low-intensity* sounds that avoid alerting their prey until it's too late to escape. Such bats, represented by the pallid bat *(Antrozous pallidus),* are said to "whisper." And as an added complication, even bats of a given species may vary their calls according to local conditions or the prey which they're hunting.

Considering that a bat's echolocation can detect the tiniest gnat or distinguish a hard-shelled beetle from a far more succulent moth, it's absurd to think that it would fly into someone's hair and entangle itself there. In fact, that's the last thing that would occur to a bat, which infinitely prefers evasion to entanglement! Perhaps this widely held myth originated with the fact that bats frequently swoop close to people's heads. However, these bats are merely homing in on insects that are attracted by *their* prey—us. The bats know precisely where they are and what they're doing, and they have not the slightest intention of lodging in our hair.

It's the combination of great maneuverability and astonishingly accurate, sensitive echolocation that makes bats in flight so fascinating to watch. Trying to capture the flight of a bat is nearly as difficult as trying to capture the pattern of a swift-flowing stream. Before the eye can register and the brain comprehend the movement, it has already changed like quicksilver. Swerving, darting, swooping, diving, changing directions almost at right angles without warning, a bat's flight is as unpredictable and indecipherable as the movements of a prestidigitator's hands. Small wonder, then, that careful observation of bats in flight is such a rewarding pastime.

In addition to older and more superstitious fears, bats are also feared because they're often portrayed as a major threat of rabies to humans. This, unfortunately, is an enormous exaggeration.

Like most other mammals, bats can and do contract rabies and can transmit it to humans. Nearly always fatal, rabies is a virus disease of the mammalian nervous system. In virtually all cases, rabies is transmitted by a bite from a rabid animal in the last stages of the disease.

Rabid animals in the final stages of the disease exhibit one of two kinds of behavior. In the "dumb" phase of the disease, a rabid animal acts extremely lethargic and may stagger and lose control of its movements. In the so-called "furious" phase, the rabid animal often attacks anything around it, biting living creatures and inanimate objects quite indiscriminately. Animals in this furious phase may also foam at the mouth. This is the behavior that most people think of in connection with a "mad dog."

At one time it was believed that bats harbored the virus without dying from it. Now scientists have learned that rabid bats do indeed die from the disease. They don't, however, exhibit the furious phase of the disease and usually don't attack other creatures. For this reason, anyone who avoids close contact with bats—especially handling them—stands virtually no chance of contracting rabies from them.

Before considering what reasonable precautions one should take to avoid bat rabies, we should first put into perspective what some regard as a great menace. From 1980 to 1996, a total of thirty-six cases of human rabies were diagnosed in the United States. Of these, twenty-one were attributed to bats.

This amounts to about 1.3 cases per year of bat-caused human rabies in the United States.

By way of comparison, attacks by nonrabid domestic dogs kill as many people in the United States in one year as bat-caused rabies does in a *decade,* and ninety-five people died of bee stings in the United States during the most recent year of reporting. Another way of viewing it is that rabies is now the second-rarest disease in the United States and Canada, trailing only polio in that regard.

From these figures, it's easy to see that we hardly need to live in terror of rabid bats. Further, having lots of bats around (excluding inside our living quarters) doesn't appear to increase the chances of acquiring rabies by more than the most minuscule degree. Why? Because hardly any of the cases of bat-caused rabies in humans have been due to our most common bats.

Only two cases of human rabies have been attributed to the big brown bat *(Eptesicus fuscus)* and none to the little brown bat *(Myotis lucifugus)*—two of our most abundant species. In contrast, the solitary silver-haired bat *(Lasionycteris noctivagans)* accounted for fifteen of the twenty-one (71 percent) of the cases of bat-caused rabies in humans.

In spite of these facts, the level of fear about rabid bats sometimes rises almost to hysteria, even among health officials who should know better. The state of New York, for example, is spending a million dollars a year "educating" the public about the dangers of rabid bats in ways that may simply exacerbate the already unreasonable fears which many people have of bats. The basis for this costly and potentially misleading campaign? The state of New York has recorded exactly *one* case of bat-transmitted rabies in its entire history!

As an example of the problems caused by the state's almost paranoid concern about rabid bats, consider the plight of a prestigious summer camp for boys. Fifty-three boys were sleeping in cabins where a bat was seen flying. Although there was no evidence that the bat was rabid or that anyone had been in contact with the bat, fifty-two of the boys had to receive the very costly rabies vaccination series (the parents of one boy refused the vaccination). At another camp, forty-four campers and counselors were vaccinated, based on health department recommendations, *merely because bats flew over them!*

In contrast, Austin, Texas, has made a virtue of its bat population. A million and a half bats have roosted under a bridge in downtown Austin for years, and large numbers of both city residents and tourists regularly gather to watch their nightly exodus, yet not a single person in the area has contracted rabies.

Unfortunately, health officials in a small minority of states (probably no more than a half-dozen) aren't the only problem. From time to time, various national and regional magazines and newspapers take up the cudgels to verbally beat on bats as a serious threat to spread rabies. A recent short piece in a national magazine is a fine example.

Citing articles in two prominent and highly respected medical journals, the author of this ill-advised piece warned that rabid bats act aggressively "just like raccoons and other infected animals. . . ." As it turns out, the so-called aggressive behavior discussed in the medical journals consisted of such things as bats biting when handled or when a sleeper rolled over onto a bat that had landed on the bed. Implying that rabid bats fly about and aggressively bite people represents a wild extrapolation of the information contained in the medical journals.

The sad part of this near hysteria is that it encourages widespread fear and killing of bats, many species of which are already in serious decline. Solid, scientifically based warnings about handling bats and the proper procedures if one has been in contact with a bat are to be encouraged. Terrifying people about bats in general is quite another matter.

It's not entirely clear, even to experts, why bat-strain rabies in humans, while extraordinarily rare, is more common than human rabies caused by bites from dogs, cats, raccoons, foxes, and other mammals. However, there is a strong suspicion that people who receive minor bites or scratches from bats may not consider them significant, whereas most people bitten by larger mammals receive rabies shots—unless, of course, the biting animal proves negative for rabies.

Although we have no reason to be apprehensive about the bats that flutter around us at night, some precautions should be followed meticulously.

1. *Never handle a bat, especially one that acts sick, unless absolutely necessary. In that event, wear leather gloves to protect against bites.* According to the organization Bat Conservation International, "Careless handling is the primary source of rabies exposure from bats."
2. *Any contact with a bat should be reported to health authorities and your physician.* They will evaluate the contact and decide whether or not rabies shots are needed.
3. *In the event that a person may unknowingly have been exposed to contact with a bat, this should promptly be reported to your state health department and your physician, as rabies shots may be warranted.* Examples of this sort of unknowing contact with a bat include such things as a bat in the room of a sleeping person or near a previously unattended child.

4. If possible, whenever contact with a bat is known or suspected, it's helpful if the bat can be brought in for rabies testing. Obviously, one should avoid touching the bat; instead, catch it in a net, paper bag, pail, or some other container.

A word about rabies shots might also be helpful. News accounts of rabies and rabid animals occasionally refer to the "painful" series of rabies shots. This view is antiquated, to say the least, and it would be tragic if anyone exposed to rabies avoided treatment for fear of painful shots.

Some years in the past, rabies prevention involved a lengthy series of shots in the abdomen that reputedly were quite painful. Now, however, post-exposure treatment for rabies consists of five shots over a twenty-eight-day period. Although this treatment is expensive, it's 100 percent effective and isn't inordinately painful.

My own experience in this regard might be helpful. We generate our own electrical power from the sun, with a backup generator. One day last summer I opened the door of the generator house and was startled when a bat flew out, brushed my arm, and landed on my pants leg.

Afraid that the bat might bite through the cloth if I frightened it, I started to take off my trousers very slowly and carefully. At that point the bat took wing and fluttered to the side of the house, where it clung to the wood. I went inside to get a bag with which to capture the bat, but it was gone by the time I returned.

I then phoned Dr. Robert Johnson, veterinarian for the Vermont Health Department. After reviewing the incident, he determined that, according to the rabies protocol established by the National Institutes of Health, I had not been exposed to rabies. However, during further discussion with Dr. Johnson, I learned about the rabies pre-immunization series.

Pre-immunization consists of a series of three shots over a period of about four weeks (a single booster shot per year is required thereafter to maintain immunity). Unlike the treatment after exposure to rabies, which provides immediate immunity, this series builds a person's immunity gradually.

I felt that this treatment was well worthwhile and went ahead with it. Each shot cost me only thirty-five dollars at a local health clinic, and I can testify to the painlessness of it; I literally didn't even feel one of the shots, and the other two were no more than tiny pin pricks. There was no soreness in my arm later, either. So much for the notion that modern rabies shots are extremely painful!

Occasionally a bat will, by some means or another, enter a home and fly about. Usually utter panic ensues, with frantic efforts to kill the creature. This is precisely the wrong reaction! Clearly, it isn't desirable to have bats in one's home, any more than mice, squirrels, and assorted other little beasties belong in our living room, bedroom, or kitchen—but there's a better and more effective way than trying to demolish the bat. In most such cases, the bat is a young one setting out on its own. It's confused, and would be just as happy to leave as the human occupants would be to see it depart. In this situation, if doors and windows are opened, the bat will usually find its own way out in a very short time.

Bats may also inhabit parts of a building outside our living quarters. It's beyond the scope of this book to list all the many ways of keeping bats from roosting in an attic, under the eaves, or in other places where they might not be wanted. Your state wildlife or natural resource agency, or an organization such as Bat Conservation International, can provide that sort of detailed information.

So far, I've tried only to demolish some of the myths, superstitions, and gross exaggerations that have been attached to bats over many centuries. Now it's time to move on and view the bat in a very different light.

The truth is that bats are enormously beneficial, especially in controlling insects. One bat can eat several thousand insects in a single night, and the 20 million Mexican free-tailed bats that roost in Bracken Cave, Texas, consume an estimated one-quarter to one-half million pounds of insects each night! Some North American bats, and a number of tropical species, are also invaluable as plant pollinators.

In fact, it's difficult to overestimate the role of bats in controlling a wide variety of insects, many of them harmful—or at least unpleasant—to humans. A single little brown bat can devour up to *twelve hundred* mosquito-sized insects in an evening. Imagine what it might be like around our homes if these industrious little insect traps weren't patrolling our surroundings night after night, snapping up mosquitoes, gnats, and other biting insects! Small wonder, then, that many people have chosen to erect bat houses near their homes to encourage a greater population of bats.

One of the more durable myths about bats is the notion that they're a sort of flying mouse. This belief no doubt made sense in the days when science was far more primitive—and error-prone—than it is now. For a long time now we've known better, however. Notwithstanding their small, furry, mouse-like appearance and their German name, *Fledermaus,* bats are not close relatives of mice at all. Indeed, the connection between mice and bats is actually so tenuous that bats are much more closely related to humans than to mice.

What *are* the bat's closest relatives? Scientific opinion varies somewhat regarding this question. The most widely accepted theory is that lemurs are the leading candidates. These small, arboreal mammals, found chiefly in Madagascar, are primitive primates. Since humans are also primates, this indeed connects our species to the bats.

Bats come in an extremely wide range of sizes. In recognition of that fact, all bats are divided into two groups, Megachiroptera and Microchiroptera. The Megachiroptera—the megabats or large bats—are fruit bats. These bats of the Old World tropics seek their food by sight rather than by echolocation. They include the world's largest bats, the so-called flying foxes; the biggest of these, a native of Java, can weigh nearly three and a half pounds, with a wingspan that may exceed six feet.

In contrast, the micro, or small, bats, which include our North American bats, are insect-eaters and hunt by echolocation. Microbats include the world's smallest bat, the bumblebee bat of Thailand. This tiny creature is well named, for its wingspan is less than five inches, and it weighs less than *four one-hundredths* of an ounce.

Our North American bats, though larger than this minuscule creature, are nonetheless extremely small. Their size is deceptive because of their large wing surface, but they weigh astonishingly little. For example, the big brown bat weighs only one-half ounce, and the little brown bat one-third to one-quarter of an ounce.

In addition to being the only mammal that truly flies, bats have other unique features. For one thing, bats roost hanging head-down. Whereas other mammals, including humans, would suffer serious ill effects if they remained upside down for any length of time, this evolutionary quirk obviously suits the bat very well. In addition to turning the world upside down for the duration of the long summer daylight hours, hibernating bats may spend weeks—even months—roosting with their heads down.

This unusual behavior is made possible by physical and chemical adaptations in the bat's body that restrict the flow of blood to the head. In turn, hanging head-down offers bats the advantage of being able to see approaching danger and to spring into instant flight. No doubt this evolutionary twist has enhanced bat survival over millions of years.

Even in their reproductive habits, bats display unique qualities. Most of our bats bear a single young per year, although some species may have two. The baby bat, known as a pup, weighs at birth a quarter of its mother's weight. The human equivalent would be a 120-pound woman giving birth to a thirty-pound baby! As an added feature found only in bats, the mother nurses her young with teats on the *sides* of her body, rather than the front.

Depending somewhat on the species, the mother bat may carry her pup, clinging to her furry skirts, on her nightly hunting forays. As the pup grows and becomes too much of a burden, the mother simply leaves it hanging in the roost. After about a month, the bat pup is able to take flight and forage for itself. Incidentally, one of the great mysteries of bat biology is how a mother can locate her pup among millions of bats in some of the largest bat caves.

Although bats have a low reproductive rate, they make up for it by a long life span; some bats have actually been known to live twenty to thirty years, a remarkably long life for such a small mammal.

Many North American bats migrate with the onset of winter weather. Some travel only relatively short distances to a cave, abandoned mine, or other suitable place to spend the winter where a fairly even temperature is maintained. Others may migrate several hundred miles to find winter quarters. Although hibernating bats need a temperature without wide fluctuations, they possess yet another unique feature to protect them during their winter rest: some hibernating bat species can actually survive with a body temperature as low as 23 degrees Fahrenheit.

There are about a thousand species of bats worldwide—about 25 percent of all known mammal species. Of these, forty-three species are native to North America. Sadly, about half of these are in severe decline, regarded either as threatened or endangered. Most of this decline is attributable to human activity. Partly it's due to carelessness or misunderstanding; such things as pesticide use, habitat destruction, development, and human disturbance—particularly during hibernation—have all taken their toll. Disturbance is especially destructive; simply agitating hibernating bats by entering their winter quarters may cause them to burn up as much as sixty-seven days' worth of energy reserves.

Even worse has been the vandalism and wanton destruction of bats and their habitat. Some people have even gone so far as to dynamite caves and abandoned mines where bats roost or hibernate, and a variety of other methods have been used to harass and kill these harmless and beneficial creatures. The bright side of this is that education is having an effect, and more and more people are coming to appreciate how useful and amazing bats truly are. This gives us at least some hope that the decline of our North American bats can be halted and perhaps even reversed—a goal we should all strive for.

7

It Isn't Acting: The Opossum

M Y T H S

- Opossums pretend to be dead when danger threatens.
- Opossums hang by their tails while they sleep.
- Young opossums travel by hanging by their tails from their mother's tail.

WE TEND TO THINK OF MARSUPIALS IN TERMS OF THE CLASSIC KANGAROO, BOUNCING ALONG WITH ITS JOEY'S HEAD VIEWING THE WORLD OVER THE TOP OF ITS MOTHER'S POUCH. Marsupials are a diverse lot, however, as demonstrated by the opossum *(Didelphis virginiana).* As our continent's sole representative of the marsupial order—creatures that carry their young in a pouch—the opossum is an object of considerable interest. Simply called "possum" most of the time, except in highly formal usage, this house-cat-sized creature has expanded its range greatly during the past few decades, bringing a new and interesting creature to many areas where it was previously unknown.

Despite their status as our only native marsupial, possums are best known for "playing possum," a phrase that has become common in our language as a term for feigning such things as illness or sleep in an attempt to deceive. Actually, the common notion of playing possum is somewhat off the mark: the possum isn't playing at all, if playing is meant as an intentional act of deception.

Possums aren't, to use a current expression, the brightest bears in the woods, and they don't think in terms of playing dead to deceive an enemy. When a possum is threatened, it's likely to first show its teeth and hiss almost like an angry cat. If that fails to frighten a would-be predator, the possum may

Opossum

run away or climb a tree. As a last resort, however, the creature falls into a sort of catatonic state, body limp and eyes wide open. This is not a conscious act of pretending, but is really a genetically programmed reflex action. Sometimes this defensive adaptation works, and a predator loses interest in a victim that appears to be dead. Even after the threat is gone, though, the possum may remain in its comatose state for hours!

The possum was once regarded as a creature of the South, coming north only as far as Virginia, Indiana, and Ohio. However, owing partly to a gradual warming of the climate and partly to transportation by humans, possums are now found in most of the lower forty-eight states, as well as in southern parts of some Canadian provinces.

I have a rather good idea just when possums first reached the area of Vermont's Champlain Valley where I grew up, mainly because of a most amusing incident that occurred in my hometown. A neighbor, Basil Muzzy, went out with his hound one evening in the mid-1950s to hunt raccoons. Soon he heard his dog barking "treed" and hastened to the spot to dispatch the coon. Instead of a coon in a tree, however, he found the dog barking at an animal lying on the ground, the likes of which he'd never seen before.

The strange creature was dead, so Basil picked it up by the tail, brought it back to his car, and dropped the deceased animal on the floor of the backseat. Then he drove to the home of a friend, who he hoped would be able to identify the animal.

When Basil summoned his friend and looked into the backseat to retrieve the beast, he found—absolutely nothing! After puzzling over this for a bit, he realized that the car windows were shut, so there was no way the animal could have escaped. Accordingly, he began turning the car inside out. He finally located the object of his search—far up behind the back of the rear seat! The animal was dead, and his friend had no idea what it was, so Basil opened the trunk, dumped in the expired creature, and closed the lid. Then he drove to the house of another friend.

Upon arriving at the home of the second friend, Basil asked for his help and opened the trunk to display his curious acquisition. Nothing! Now, the trunk had no exits and contained only a spare tire without a tube in it, so it wasn't difficult to fathom where the strange animal had gone. Basil felt around inside the tire and soon came upon the missing mammal, which he promptly hauled forth. The second friend didn't know what it was, either, but it was obvious the creature was dead.

About that time, Basil remembered the phrase "playing possum" and began to suspect the truth. He next drove to the home of the local game warden, who quickly confirmed that the stranger was indeed a possum—seemingly dead, but still very much alive—the first one seen in those parts.

The name *opossum* is one of those adapted without much change from the Native Americans; it comes from the Algonquian *apasum,* "white animal," and most possums, especially those farther north, are indeed a very light gray—sometimes almost white. In 1612, Captain John Smith of Pocahontas fame was the first European to send back word of this novel creature, which he described in these terms: "An Opassum hath a head like a Swine, & a taile like a Rat, and is of the Bignes of a Cat." Captain John was evidently a keen observer, for this is a remarkably apt, if somewhat superficial, description.

Aside from a very long, pointed snout with fifty teeth, the possum's most notable feature is its long, naked, ratlike tail. This useful appendage is prehensile, and possums are widely depicted as hanging by their tails while sleeping. This is a myth: biologists who work with possums have never seen this behavior. Possums do, however, wrap their prehensile tails around branches in order to brace themselves and steady their position in trees, using the tail almost as a third hand, and one might conceivably suspend itself by its tail very briefly. Possums don't, however, hang by their tails for any length of time, and they certainly don't sleep in that fashion.

If possums don't dangle by their tails, they assuredly hang by their hind feet. This behavior is made possible by a most curious adaptation that helps them clamber about in trees, gripping branches securely. Four of the toes on each hind foot have sharp claws, but the fifth, the big toe, lacks a nail and is opposable, like a human thumb. This odd-looking digit, which appears almost as if its tip had been amputated, enables the possum to grip branches with such dexterity that it can use its front feet to gather fruit while suspended by its hind feet.

True omnivores, possums will eat virtually anything. In fact, those who know the possum well often describe it as the quintessential omnivore—a sort of walking garbage can. Possums have a reputation in story, folklore, and song for loving persimmons when the puckery fruits ripen to sweetness in the fall. True, possums do devour persimmons with gusto, but they also feed avidly on other fruit, birds' eggs, mice, slugs, earthworms, nuts, snakes, garbage, lizards, carrion—the list goes on and on, encompassing almost everything imaginable.

Possums also seem to relish cat food. My aunt, a great cat lover, used to feed several semi-wild cats on the back porch of her Vermont farmhouse. This small porch was enclosed, but she left the door open to give the cats ready access to the tray of cat food that she regularly placed there. When my aunt opened the door to the porch one day, she was startled to see a strange creature intermingled with three or four cats, the whole crew industriously putting away the canned cat food while studiously ignoring each other.

She quickly deduced that the uninvited guest was a possum, and was a bit wary of it at first. The possum, however, continued to show up now and then, although not on a daily basis, to eat cat food, so my aunt soon became accustomed to its presence. Considering the possum's willingness—even eagerness—to devour anything even remotely digestible, its predilection for cat food isn't to be wondered at. Far more surprising was the apparent tolerance of cats and possum for each other. Perhaps, being of roughly equal size, neither felt threatened by the other.

Aside from being a marsupial, the central feature of possum biology is the creature's life span. For its size, the possum is one of the shortest-lived animals in the world. Most mammals of house-cat size see a portion of their numbers live for at least three to five years, and frequently considerably more. House cats, for example, routinely live into their teens, and a few even manage to be around for a full two decades. Possums, in contrast, almost never live to be more than two years old, and most fail to make it even that far. Why is anybody's guess. Most wild animals in captivity live considerably longer than their wild compatriots, but possums are an exception; even captive specimens rarely make it past the age of three. Wild or captive, they apparently live life in the fast lane and simply run down after two years or less.

Possums have enemies, of course. Perhaps the major cause of possum mortality, at least in more heavily settled areas, is the automobile. Possums aren't swift when they cross highways, and often feed on roadkill, thereby exposing themselves to the same agent that did in the roadkill. Possums are also widely hunted in rural areas for their meat and fur. Still, human predation, intentional or otherwise, seems to have little impact on the possum population. So brief is the possum's life span that much of this mortality may only substitute for other early causes of possum demise.

Much the same can probably be said of other enemies. Owls, snakes, dogs, coyotes, and assorted other predators take their toll. So do disease and parasites. None of this seems to matter much, though. Possums quickly wind down and expire, even if something else doesn't get them first.

Considering the possum's exceptionally brief life, how has it managed to survive for millions of years? Its lineage is an ancient one; its marsupial ancestors date back at least 85 million years, well into the age of dinosaurs, and our present day possum split off from another opossum as far back as 75,000 years ago.

The secret of the possum's survival seems to be its exceptional fecundity. A female possum bears her first litter when she's only six to nine months old. If she lives long enough, she'll have a second litter, but only a very few possums live long enough to have a third. A litter often consists of ten to fifteen young—sometimes more—and averages about nine. This combination of very early reproduction and huge litters has effectively ensured the survival of this creature and its ancestors.

As with other marsupials, possum young are born virtually in the embryonic stage. This is hardly surprising, considering that their gestation is an astonishingly brief thirteen days! The tiny newborns are only about the size of a raisin, and a whole litter can be fitted into a tablespoon. The most critical moments for a newborn possum occur when the tiny, larvalike creature must make its way about two inches to the mother's pouch. Although most succeed, some don't, and the latter simply die. Those that reach the pouch have an excellent chance of survival as, warm and protected, they find a teat and begin feeding. There they remain, firmly attached, for several weeks.

Possums are nocturnal, and the mother travels widely on her nightly feeding expeditions. Meanwhile, the babies go with her, riding in her pouch for about three months. This is an extremely useful evolutionary adaptation, for the mother never has to worry about returning to a den to feed her young. As a result, she can range far and wide in search of food, denning here and there for a rest as the spirit moves her.

After this three-month interval, the young possums emerge from the pouch and for a few days remain in a den while the mother forages. Thereafter, the babies resume traveling with their mother. Possum babies are often portrayed dangling from the mother's tail by their own, but, although it makes an appealing picture, this is just another myth. In real life they either scurry along beside their mother or cling to her long fur.

In just a few more weeks the young possums are grown and set out on their own. After several additional weeks they'll produce another generation of possums, thereby ensuring that this unusual creature will survive, despite predators and an abnormally brief life span.

8

Like Knights of Old: The Nine-banded Armadillo

MYTHS

- The armadillo's main defense is to curl into a tight ball.
- Armadillos feed heavily on the eggs of quail, turkeys, and chickens.

MEDIEVAL ARMORERS THOUGHT THEMSELVES VERY CLEVER WHEN THEY DEVELOPED A METAL CASING OF FLEXIBLE PLATES FOR MOUNTED KNIGHTS, WHO THEN WENT ABOUT SLAYING DRAGONS, RESCUING FAIR DAMSELS IN DISTRESS, AND HAPPILY DISPATCHING EACH OTHER WITH LANCE, BATTLE-AX, AND MACE. However, armadillos and their ancestors perfected a nearly identical, and far more practical, system millions of years before; they had no need to depend on armorers, for they simply grew their own armor! Indeed, the name *armadillo,* a gift from Spanish explorers and conquistadors, means "little armored one"—a good name for this odd and intriguing little creature.

Most species of armadillo are tropical or subtropical, but the nine-banded armadillo *(Dasypus novemcinctus)* has gradually migrated northward through Mexico and into the United States. It now inhabits Louisiana, most of Texas, much of Oklahoma and Arkansas, and parts of Kansas, Missouri, Mississippi, Alabama, Georgia, and Florida.

Our North American armadillo is roughly the size of a house cat. The head and body combined are about sixteen inches long, while the slender, pointed tail is an inch or two shorter. Adult females weigh from eight to thirteen pounds, while the slightly larger males can reach seventeen pounds.

As its species name *novemcinctus* (from Latin *novem,* nine, and *cinctus,* that which girds) indicates, our North American armadillo has nine plates, or bands, sandwiched between a large front plate and a large rear one. These nar-

Nine-banded armadillo

row middle plates allow the armadillo to be quite flexible, a quality it would be totally lacking if it had only two or three large plates. The head and tail are also armor-plated, as are the legs. Only the underside lacks protection.

It's commonly thought that this strange little animal's first line of defense is to curl into a tight, armored ball in order to protect its soft underside. However, the armadillo much prefers to run into brush so thick that its enemies can't follow, dive down a burrow, dig its way swiftly out of sight, or even escape by swimming. Only as a last resort does it curl into an armored ball, from whence it can kick an attacker with powerful hind legs and long, strong claws.

Although an armadillo's tough, horny plates do serve *in extremis* as a defense against predators, they probably function more as a shield against thorns, spines, and sharp twigs. The armadillo mostly inhabits areas where sharp, spiny plants seem to be the rule, and its armor enables the little creature to escape danger by scurrying with impunity into thick, thorny brush too daunting for most predators to tackle.

One of the armadillo's characteristics, however, represents a distinct defensive disadvantage in the modern world: when startled by the sudden approach of something like a speeding auto, the armadillo tends to leap into the air to about bumper height. There, despite the little animal's armor, the car always wins!

The subject of road-killed armadillos leads to the story of a classic practical joke. A former colleague of mine, driving back to his home in the Midwest from a conference in Texas, picked up a number of freshly killed armadillos along the highway. Then, when he was well north of the armadillo's range, he began to surreptitiously drop his defunct passengers at strategic points along the highway where they would be highly visible. As a result, great excitement and puzzlement ensued among wildlife officials and the news media in several states because of what seemed a sudden, unexplained incursion of the little armored ones! As far as I know, the perpetrator of this brilliant deception never told anyone but his fellow workers about it.

Armadillos, along with anteaters and sloths, belong to the order Edentata, which means "without teeth." Although some members of this order, such as the anteater, are literally toothless, the armadillo does have twenty-eight to thirty-two (usually thirty-two) very primitive teeth, in the form of simple pegs. Not surprisingly, then, its omnivorous eating habits concentrate on foods that require little chewing. Worms, insects, grubs, and other small invertebrates make up a large part of the armadillo's diet. It also has a long, sticky tongue, somewhat like an anteater's, which is handy for dredging up ants and termites. Soft fruits, berries, birds' eggs, and carrion also find their way down the armadillo's rather unfussy gullet.

Because so much of the armadillo's food comes from grubs, worms, ants, and other subterranean invertebrates, the right kind of soil is a critical factor in armadillo habitat. Soft soil will support a denser population of armadillos than harder soil, and if the soil in an area is excessively hard, armadillos won't live there.

In some areas of the United States and Mexico, armadillos are regarded as a serious threat to the nests of quail, turkeys, and chickens. This appears to be a bum rap that isn't supported by the evidence. One study, for example, found the remains of eggs in only five of 281 armadillo stomachs. It appears likely that armadillos are simply being blamed for the depredations of other creatures.

As already noted, the armadillo is a master digger. If the ground is reasonably soft, it can dig its way out of sight in a just a minute or so. This remarkable aptitude for speedy excavation is due to the armadillo's feet, which are admirably equipped for the task. The middle toes—two of the four on the front feet and three of the five on the rear feet—are the longest and are tipped with formidable claws.

Because armadillos have only a few vestigial hairs on their bodies, they are not well insulated against either heat or cold. When the weather is hot, armadillos are mostly active from evening until dawn; in cold weather, they're mainly active during the warmest part of the afternoon, and they can't survive in areas subject to prolonged spells of subfreezing weather.

When the weather is either too cold or too hot, armadillos spend much of their time in dens. These are mostly burrows in the ground as much as fifteen feet long, although they also use dens in the rocks. Great diggers that they are, armadillos usually have numerous burrows. In hot, dry country, they often congregate around streams and water holes, where they take cooling mud baths.

A suit of armor is by no means the armadillo's only unusual feature. This oddity among mammals is full of surprises, and its reproductive style is every bit as bizarre as its outward appearance. For openers, armadillo copulation itself is decidedly different from that of most mammals. Although the creature's nine middle plates give it a good deal of flexibility, it isn't sufficient for the standard "mount from the rear" approach. As an added complication, its genitals are located underneath. Undaunted, armadillos surmount this obstacle with ease by having the female turn on her back during mating.

The armadillo's aberrant traits continue after breeding. Mating mostly occurs in July, but the fertilized egg from a July mating doesn't attach to the uterus until November. This phenomenon, known as *delayed implantation,* also occurs in members of the weasel family and among bears (for a detailed description of its advantages, see chapter 14), but it's far from the mammalian norm.

But wait—the reproductive ways of the armadillo become even more bizarre! The single fertilized egg divides into four embryos; after implantation, these share the same placenta and grow into identical quadruplets of the same sex. Thus an armadillo always bears either four male or four female young. Following a gestation of about 120 days after implantation, these are usually born in March in a burrow with an enlarged chamber. This chamber, which serves as both birthing room and nursery, is filled with soft plant material such as grass and leaves.

The baby armadillos look exactly like miniature adults. They can walk after only a few hours and begin to follow their mother on her rounds in just a few weeks. Although they nurse for only about two months, they continue to stay with their mother for a few additional weeks before going off to live on their own. They'll be mature by the next summer and will breed at that time.

Armadillo oddities seem to have no end. Consider the creature's behavior when it encounters a stream. Although, as might be expected, an armadillo lacks buoyancy because of its armor, one can hold its breath for up to six minutes—an adaptation that evidently helps it avoid inhaling dirt and dust while digging burrows. When an armadillo comes to a stream that isn't excessively wide, it simply wades in and walks along the bottom to the far side. But what of wider waterways? That's no problem: the versatile armadillo just sucks in air and inflates its digestive tract. Then, with this extra buoyancy, it swims across with ease. Medieval knights in full armor undoubtedly fared far worse than this resourceful little creature when they were pitched off their chargers into deep water!

Both medieval knights and Spanish conquistadors no doubt boasted of their distinguished ancestry, but armadillos can claim a vastly longer lineage. The earliest known ancestors of today's armadillos date back some 55 million years—only about 10 million years after the demise of the dinosaurs. In time, quite a diverse array of armadillos and their close relatives evolved, some of them quite astonishing.

For example, within the past million years, during the Pleistocene Epoch, there was an enormous armadillo called *Chlamytherium.* This great creature, which was herbivorous, had the bulk of a rhinoceros. Far more recently the so-called giant armadillo *(Holmesina septrionalis)* inhabited Florida. Although far smaller than *Chlamytherium,* this armadillo was nonetheless huge compared with our present-day specimens, for it was six feet long and weighed an estimated six hundred pounds. There were humans in Florida by eleven thousand years ago, and the giant armadillo survived for at least another twelve hundred years, so early humans must have actually encountered this hulking fellow.

Perhaps even more fascinating than these ancient armadillos is a branch of their family tree known as *glyptodonts*. Also known as "turtle armadillos" because of their superficial resemblance to turtles, glyptodonts had a carapace, or single large plate, that covered their entire upper side, making them just as inflexible as turtles. In turtle fashion, they also could retract head and neck inside the protective shell.

One of these glyptodonts was truly gigantic: it was fourteen feet long and towered an astounding fifteen feet high! Despite its vast bulk, however, this perhaps wasn't the most bizarre of the glyptodonts. That prize probably belongs to the creature known as *Doedicurus*.

Doedicurus must have presented an amazing sight. Along with its inflexible shell, it had a long, stout tail that tapered outward and terminated in a large, spiked ball. No doubt this creature lashed its enemies with this potent weapon, much as knights of yore clobbered each other with maces!

Present-day armadillos are valued for their meat in many areas, although they probably aren't as widely hunted now as they were in the past. I once had the opportunity to try armadillo meat at a dinner in Texas; it was delicious, tender and moist, with a flavor similar to that of pork. The hides are also used to make a variety of items for sale, mostly in the tourist trade.

Many people have a virtual love affair with the armadillo, especially in Texas, where it's the official state mammal. Armadillo races are held, and so are armadillo "beauty pageants." For the latter, proud owners groom their armored charges to ensure that they look their best. Beauty is indeed in the eye of the beholder!

Not everyone takes such a benign view of armadillos, however. Owing to their propensity for digging up grubs and earthworms, the little beasts can be the bane of homeowners who take pride in their lawns. Even worse can be their effect on a golf course, where an armadillo can gouge up a lot of valuable turf in a short time.

Many armadillo oddities have already been described, but nothing about them is stranger than their role in Hansen's disease, more commonly known as leprosy. According to Dr. Richard Truman, Chief of Microbiology at the G. W. Long Hansen's Disease Center, the armadillo is the *only* mammal besides humans that's known to develop leprosy at a high level of frequency in a natural population. Chimpanzees and sooty mangabys have been known to contract the disease, but these cases have been rare. On the other hand, up to 20 percent of wild armadillos in some areas harbor the bacteria that cause leprosy.

At present, says Dr. Truman, there is no proof that humans can acquire Hansen's disease by handling infected armadillos, nor is there any way of assessing the level of risk involved. There is an immense amount of human contact with infected armadillos, considering the number of people who adopt wild armadillos as pets, pick them up to move them out of yard or garden, hunt them for meat, or handle road-killed specimens. Despite this level of contact, only thirty to forty new cases of the disease annually are diagnosed in the United States that don't result from infection in other parts of the world, so the level of risk obviously is not great. Nonetheless, advises Dr. Truman, people should at least take this information into account before deciding whether or not to handle armadillos.

If, however, armadillos present a possible risk of acquiring Hansen's disease, they might ultimately be of use in finding a vaccine against this ancient scourge. One of the greatest difficulties in research on Hansen's disease is the fact that *Mycobacterium leprae,* the cause of the disease, can't be grown successfully except *in situ,* that is, in or on a living organism. This feat was first successfully accomplished on the foot pads of laboratory mice in 1961, and it was theorized that a slightly lower temperature outside the body might have made this possible.

Enter friend armadillo, which, among its many other unique characteristics, has an internal body temperature 2 degrees to 3 degrees Celsius lower than that of humans. Researchers found in 1968 that *Mycobacterium leprae* would, in fact, develop inside armadillos injected with it. Thanks in part to subsequent research, there are currently three studies testing a vaccine against leprosy, although it's too early to tell whether or not it will be effective.

Armadillos have had a long evolutionary journey and have stood the test of time very well. Although they've been food for a variety of predators, including humans, their various defenses have been more than adequate to the task of preserving the species. Their worst enemy today is the automobile, but the armadillo's reproductive rate seems more than adequate to compensate for this loss. In fact, this strange little creature is actually increasing its range and appears to be on its way to surviving for many millennia more.

Eastern (red spotted) newt—red eft stage

9

One and the Same: The Newt and the Red Eft

MYTHS

- The newt and the red eft are two different species.

PEER INTO PONDS, POOLS, AND QUIET STREAMS IN THE EASTERN UNITED STATES AND CANADA, AND YOU MAY SEE SMALL, BROWNISH, RED-SPOTTED CREATURES THAT BEAR A PASSING RESEMBLANCE TO MINUSCULE ALLIGATORS. They bear no relationship to alligators and other reptiles, however; instead, they're a type of salamander known as the newt—in this case, the Eastern or red-spotted newt *(Notophthalmus viridescens)*.

Like frogs and toads, salamanders are amphibians, meaning that they can lead a double life, one stage aquatic and the other terrestrial. The newts, however, have added a unique extra complexity in that they have not two but *three* distinct life stages, two of them aquatic and one terrestrial. Further, in a rather mystifying anomaly, some red-spotted newts skip the second, intermediate stage and spend both life stages in the water!

When is a newt not a newt? The answer to this riddle is linguistic rather than biological, and lies far in the past. In Old English (also termed Anglo-Saxon), spoken in England from about A.D. 450 to A.D. 1050, this type of salamander was originally known as an *efete*. This soon became *evet* and, in an unusual shift from *v* to *w* a little later still, *ewt*. This brings us into Middle English, which reigned from just before the Norman Conquest until about 1475.

But a funny thing happened to the ewt on the road from Middle English to Modern English. The articles *a* and *an* had sometimes been combined with the words which followed them. Then, late in the Middle English period, they began to be separated again, and considerable confusion was the result. In this process, *an ewt* (or *anewt*) became *a newt*. This uncertainty and subsequent incorrect separation worked both ways. Old English *naedre,* for example, gave way to Middle English *nadder,* which was finally transformed from *a nadder* (or *anadder*) to *an adder.* In similar fashion, *a napron* became *an apron.* Of such vagaries are names sometimes fashioned! This shift had clearly taken place before Shakespeare's time, for "eye of newt" is one of the ingredients in the infamous witches' brew in *Macbeth.*

Newts are common and widely distributed from the Canadian Maritimes to southern Ontario and south to eastern Texas. The adult newt is not very large—three and a half to almost four inches long. Its color varies somewhat from yellowish to greenish brown on the upper side, with scattered black dots and larger, black-ringed red spots along the back. The underside is colored with yellow in varying shades, sprinkled generously with tiny black dots. The tail, which constitutes about 40 percent of the newt's total length, is keeled—that is, there is a thin, soft ridge running the length of the tail on both the upper and lower sides.

Life for the red-spotted salamander begins during the spring in quiet water, sometimes in a pool or pond that eventually dries up in late summer, although not early enough to kill the newt's completely aquatic larval stage. The advantage to the newt of such temporary pools is that they contain no fish, which are major predators of both larval and adult newts in permanent waters. Adult newts, though mostly aquatic, are capable of migrating across land to reach permanent pools if their more temporary breeding pools have dried up.

After mating, the female newt lays her eggs, which are attached to the stems or leaves of either aquatic or temporarily submerged vegetation. For this reason, newts greatly prefer waters that contain abundant plants, since these serve both to hold the eggs and to protect the adults from predators. There's a wide variation in the number of eggs laid by female newts; some lay fewer than a hundred, while others may deposit nearly five hundred.

Depending somewhat on water temperature, the eggs hatch in about a month. The tiny larvae have keeled tails and feathery external gills located just behind the head. This larval stage is exclusively aquatic, totally dependent on water to provide it with oxygen and sustenance. Newts in all stages are carnivorous, and the larvae eat a wide variety of small fare. These include tiny crustaceans and snails, larvae of insects such as mosquitoes, and water fleas.

The newt larvae continue to grow and develop throughout the summer. Then, just before the onset of autumn, metamorphosis takes place. The gills gradually disappear, the tail loses its keels, and the smooth, slippery skin becomes rougher. At last the young newt is ready to enter the second stage of its life and leave the water for dry land.

Now let's switch to an entirely different scene. Imagine going for a walk in the summer woods during or just after a rain. There you're likely to encounter small creatures that, like the newt, look somewhat like tiny alligators; in fact, they bear considerable resemblance to a newt in size and shape, yet they lack the keels on the tail and are colored red or orange. This is the red eft, which changed from drab brown to red or orange just before it left its aquatic habitat.

Who would think, looking at this often brightly colored forest dweller, that it was the rather drab aquatic newt? Logically, one would assume it was an entirely different species, but in this instance logic would be wrong. After its gills disappear and its lungs develop, the young newt, now officially an eft, or red eft, usually waits for a rainy night in late summer or early autumn. Then, forsaking the aquatic world of its natal pond or stream, it sets forth to seek nearby forest land.

Once ensconced in its terrestrial habitat, an eft spends much of its time hiding in cool, damp places—beneath logs or sticks, under leaves, burrowed under rocks, and similar places. From such places of concealment, the eft emerges at night to feed on the smaller sorts of spiders, snails, worms, caterpillars, and other invertebrates. On dark, rainy days, however, efts emerge in daylight hours to forage and wander about. The number of them at such times is quite astonishing, considering that a walk through the same area before the rain would reveal nary an eft!

The color of the eft's dry, slightly pebbly skin is quite variable. Many efts are a bright red that stands out like a beacon against the rather dull colors of the forest floor. Others are an orange that's nearly as bright, but still others are of varying lesser intensities of red and orange, all the way down to a sort of dusky reddish brown. In common with the adult newt, however, they all have little black spots and the larger black-ringed red spots along both sides of the back.

At first glance, the bright color of the efts might seem to work to their disadvantage by making them more visible to predators, but efts are to some degree toxic—though not for humans to handle—and scientists speculate that their bright color may serve as a warning to predators that they should be left alone.

The time that efts spend in this intermediate, terrestrial stage is as variable as their color. At least two years, but sometimes as many as seven, elapse before they begin to assume the form of the adult newt. This transformation takes place over several months, and the efts, now sexually mature, migrate back to water. There they develop the fully adult form, including the keeled tail. This migration back to water can occur any time during the warmer months.

The eft can be regarded as a wonderful adaptation that greatly increases the chances of the newt's survival. Because it's terrestrial, the eft stage allows newts to breed in ponds that at least occasionally dry up by late summer or early autumn, thereby eliminating fish, one of their most important predators. Even if waters that are generally more permanent dry up in a lengthy and extreme drought, the efts will survive to return to water somewhere and perpetuate the species.

The aquatic adult newts feed on all sorts of insect larvae, especially those of mosquitoes and midges. Thus they're valuable from a human perspective by controlling many ferocious little insects that love to sip our blood. Newts also eat a wide variety of insect nymphs, adult insects, snails, eggs of frogs and toads, and even the larvae of their own species.

In winter, adult newts hibernate either underwater or on land beneath objects such as logs. However, some remain active all winter beneath the ice. Efts also hibernate during their years on land. They burrow beneath the leaves or other litter on the forest floor, or crawl beneath logs and other objects that offer shelter for the winter.

This account doesn't quite complete the tale of this rather strange and very interesting little creature, however. As previously mentioned, some newts skip the eft stage entirely. These individuals, known as *neotenes,* pass directly from the larval to the adult stage without ever leaving the water. In the process, they retain such larval characteristics as external gills, although these may be substantially smaller than the gills on a normal newt larva. It's uncertain why some newts become neotenes—one more mystery in the incredibly complex world of nature.

Although newts are fortunately a very long way from being an endangered species, they can suffer locally from an increasing number of highways that interfere with the outward migration of efts and their return to water as adults. On rainy nights, large numbers of efts or newts may be killed by cars while crossing the highways. Drainage of wetlands and pollution can also pose threats to local populations. On the brighter side, construction of many farm and recreational ponds has created new habitat for newts and other amphibians, which should help to ensure the future of these interesting and enjoyable little creatures.

10

The Lowly One: The Common or American Toad

MYTHS

- Handling a toad can give you warts.
- Toads are ugly, nasty creatures.
- A toad urinates when it's picked up.

HOW SHALL I LOATHE THEE, LET ME COUNT THE WAYS. Pity the poor toad. Down through the ages this inoffensive and beneficial little creature has been regarded at best as downright ugly and at worst as vile and loathsome. Even frogs have fared far better: they, at least, have been accorded the possibility of reconversion to handsome princehood via the kiss of a benevolent and presumably beautiful princess. Alas, no similarly pleasant fate has ever been recorded for a toad! Even Kenneth Grahame, in his classic *The Wind in the Willows,* portrays Mr. Toad as a bumbling, conceited ass, albeit a well-meaning and generally likable one.

The much-maligned toad has even been associated with evil and the demonic. In the first scene of Shakespeare's *Macbeth,* for instance, the witches depart when they say, "Paddock [a term for the toad] calls. . . ." Then, in the famous cauldron scene, in which the witches brew their "hell-broth," they throw into the potion "Toad that under cold stone, days and nights has thirty-one. . . ." And in *As You Like It* we find the phrase "like the toad, ugly and venomous."

Nor was Shakespeare the only one to defame the unfortunate toad. For instance, the poet Philip Larkin penned the line, "Why should I let the toad *work* squat on my life?" Even in everyday language we indirectly heap scorn on the little creature: a *toady* is defined as a cringing, servile person, and a despised individual may be referred to contemptuously as a toad.

Common (American) toad

In speaking of "toads" here, only three species are involved. Shakespeare was undoubtedly writing about the common toad of Great Britain *(Bufo bufo)*, and this chapter is mostly devoted to the common or American toad *(Bufo americanus)*, which is nearly identical to the British common toad. Woodhouse's toad *(Bufo woodhousii)*, also known in the West as the "common toad," is likewise quite similar to these two. There are many other toads, however—approximately three hundred species worldwide and eighteen species in North America—a number of them vastly different from the common toad. Many of the North American species are in the Southwest, but the misnamed horned toad isn't one of them; rather, it's a little lizard, spiny-looking but harmless.

As if the nasty things already cited weren't sufficiently defamatory, further indignity has been heaped on the toad by accusing it of causing warts—a folk superstition that many still believe. Toads, of course, have nothing whatever to do with human warts, which are caused by a virus. The toad's so-called warts are not warts at all, but glands that have a defensive purpose. These glands, especially the two large ones behind the eyes, give off a slightly poisonous substance that makes the toad quite unpalatable to many predators.

This poison is far too mild to harm humans who handle toads, although it might sting a little if it gets into a person's eyes. Therefore, it's advisable to wash one's hands after handling a toad—but not because of any danger of warts! Incidentally, those two large glands behind the toad's eyes are sometimes mistakenly referred to as parotid glands. Parotid glands are salivary glands in mammals; the toad's two prominent glands are properly called *parotoid* glands (pronounced *pah-ROH-toid.*)

If one examines a toad carefully, with an open mind, the creature ceases to be ugly and exhibits some rather attractive characteristics. On closer inspection, its seemingly drab exterior becomes a handsome mosaic of colors on the upper side, ranging from tan to terra cotta to rich, dark browns, while black and shades of gray lend further variety to this pattern. The underparts, by contrast, are a nice cream color with some black spotting. In addition, the toad's gold-rimmed eyes are very lovely. Despite its squat shape and warty appearance, the common toad is far from ugly!

Centuries of bad publicity notwithstanding, toads are extremely useful creatures. They consume large quantities of insects, slugs, and other creatures that can seriously damage vegetables and other desirable plants. Therefore, toads should be considered a welcome addition to anyone's grounds or garden, and encouraged in any way possible.

Fortunately, the toad's sterling qualities are finally being recognized. In England, for example, there is Madingly Toad Rescue, an organization named for the village of Madingly. Owing to heavy traffic, toads were being killed wholesale as they crossed highways in an attempt to get to or return from their breeding ponds. To save dwindling toad populations, Madingly Toad Rescue was formed by concerned citizens.

With experience, this organization has developed several effective techniques for saving toads. For instance, construction of amphibian tunnels under the highways has worked very well. However, the most successful method has been for volunteers to pick up the migrating toads, put them in buckets, and move them to the other side of the highway. Volunteers have also been aided in this endeavor by temporary net fencing, which funnels and concentrates the toads so that they can be located and picked up more easily.

While appreciation of toads tends to be a bit less dramatic on this side of the Atlantic, gardeners and homeowners have increasingly come to understand the benefits of a healthy toad population. Indeed, modern gardening books often emphasize the value of toads, and gardening supply catalogs feature little toad shelters that can be placed in gardens and around homes to attract these helpful tenants.

Toads are amphibians, with all that the term implies. The word is derived from the Greek *amphi,* of two kinds, and *bios,* life, and refers to the fact that toads and other amphibians can live both on land and in the water. Indeed, water or a very moist environment is required for amphibians to reproduce, although many spend most of their lives on dry land.

Amphibians are a truly ancient race. They alone, among primitive vertebrates, first crept out of the primal waters to set foot on land some 360 million years ago—an almost unimaginable span of time. This was at the beginning of the Carboniferous period, when plants such as the giant tree ferns flourished in immense profusion, eventually to be transmuted into coal beds by the labors of eons.

This was no small step for life on our planet. Up to that time, all vertebrates were aquatic; this transition, which might be termed the Great Leap for vertebrates, was truly monumental, because it paved the way for the development of all vertebrate life on earth. Radical evolutionary changes were vital in order to make this momentous transition from aquatic to terrestrial environment. The ability to obtain oxygen from the air was critical; this required the development of lungs and moist skin that could absorb oxygen. Limbs that could support these new creatures on land and enable them to move about were also necessary.

Without other vertebrate predators, at least for a very long time, these early amphibians were able to prosper and evolve further. Still, this evolution was incredibly slow in terms of our human time frame. The earliest known ancestor of toads and frogs, barely recognizable as a very primitive prototype, dates back to the early part of the Triassic period, about 240 million years ago, or 120 million years after the first amphibians appeared. A mere 50 million years or so later, in the early Jurassic, evolution had proceeded apace and produced what could reasonably be termed the first recognizable frog. To put this development in perspective, the earliest true frog evolved even before most types of dinosaurs!

Ancient times were the heyday of amphibians, which comprised perhaps fifteen major groups at their peak. However, the development of reptiles, followed by birds and mammals, provided serious competition, and amphibians slowly declined. Today there are just three remaining orders of amphibians—frogs and toads, salamanders, and caecilians, a group of wormlike tropical creatures.

In a sense, amphibians represent a sort of arrested development. Unlike the reptiles—which, incidentally, evolved from an amphibian ancestor roughly 315 million years ago—amphibians never took that final step to become completely terrestrial. Although many reptiles, such as turtles and crocodilians, spend most of their time in the water, all reptiles (except for some sea snakes) bear their young on land, and many are completely terrestrial.

Reptiles were able to accomplish this transition by developing eggs that wouldn't dry up on land and skins that greatly reduced the loss of water. In contrast, the soft, gelatinous amphibian eggs are useless out of water, and amphibians themselves can only withstand hot, dry conditions by protecting themselves in a moist environment. Perhaps the enormous early success of amphibians was a mixed blessing, halting further development so that amphibians remained in a sort of evolutionary limbo, neither fish nor fowl, so to speak.

The question sometimes arises, what is the difference between frogs and toads? From a strictly scientific point of view, there's hardly any difference. Frogs have teeth in their upper jaw, whereas toads lack them, but otherwise the distinction between toads and frogs is largely an artificial one for purposes of convenience. *Most* toads have dry, rough skin, a plump body, and short legs best suited for hopping. *Most* frogs, on the other hand, have smooth, moist skin, a fairly slender body, and long legs well suited for leaping. However, there are frogs with rough skin and toads with smooth skin, for example, so frogs and toads can't be differentiated simply by appearances.

There has been a great deal of concern among scientists because of a recent worldwide decline in some species of frogs and toads. Various theories have been propounded for this decline, ranging from pollution to an increase in ultraviolet rays from sunlight, but no definite conclusions have been reached. Fortunately, the common toad seems to be maintaining its numbers.

Life for toads (or frogs) begins in the water. After emerging from hibernation in March or April—or even in May, depending on the climate—male toads make their way toward suitable breeding areas, such as shallow marshes, ponds, and pools. There the males begin to sing, a loud, high trill that can continue uninterrupted for as much as thirty seconds. This sound is produced when the male inflates his vocal sac and vibrates it during the course of the call. The sound of trilling toads at a discreet distance is a welcome sign of spring, but a number of males trilling simultaneously at very close range can be almost deafening!

This trill evidently serves to attract female toads and perhaps gives them an indication which male might make the best mate. Mating takes place when the smaller male toad climbs on the female's back and uses his front legs to hold her tightly in a grasp known as *amplexus*. During breeding, males have horny pads on the first two toes of each front foot to aid them in holding the female.

The female lays her eggs during amplexus; unlike frogs, which lay their eggs in a large clump, the toad's eggs are encased in two long, slender strings of gelatinous material. These strings can exceed sixty feet in length and contain up to fifteen thousand eggs! Meanwhile, the male fertilizes the eggs as they slowly emerge from the female.

Male toads aren't overly endowed with brains when it comes to mating. They often try to mate with almost anything that floats and moves, such as sticks, leaves, and even other male toads. Indeed, a male toad suddenly squeezed in amplexus by another male often (and justifiably) squeaks indignantly until the offender releases him. Eventually the eager but misguided male lets go of stick, leaf, or other male and moves on until he successfully locates a female.

Whenever submerged plants are available, toads prefer to weave their egg strings among them, although the eggs aren't actually attached to the vegetation. We've often observed this behavior in our pond. First, the mating pair, with the male on top gripping the much larger female in amplexus, appears on the surface. After breathing for a brief period, the two dive to reach underwater growth at a depth of two or three feet. Still plainly visible in the pond's clear water, the couple moves slowly among the plants while continuously laying down the two strings of fertilized eggs. After perhaps five minutes, the toads drift to the surface to breathe again, with the trailing egg strings still attached to the female. This cycle is repeated over and over until egg-laying is done and the two toads separate.

Depending on water temperature, the eggs hatch in anywhere from three to twelve days (the warmer the water, the shorter the time until hatching). The tadpoles, often humorously called "toadpoles," can easily be distinguished from frog tadpoles because they're black. They often move in schools, and feed hungrily on algae and minuscule bits of organic debris, all the while growing steadily. Just as steadily, a variety of predators take their toll; these include dragonfly larvae, predaceous diving beetles, fish, and leeches.

When they are just under a half-inch long, the "toadpoles" begin their metamorphosis. As with other frogs and toads, the tadpoles begin to absorb their tails and develop legs. This process continues for approximately two months. At the end of that time, the young are fully formed little toads, ready to leave the water for the second part of their dual lives.

When the young toads begin their migration onto land, they present quite a spectacle. Only about a third of an inch long, they swarm up onto the land by the hundreds, so small that at first glance they may be confused with insects. Under normal conditions, the toadlets disperse rapidly, heading off in all directions to seek suitable habitat. Under drought conditions, however, they may gather in moist places to await better conditions for dispersal.

A young toad sheds its skin every few weeks while it's growing rapidly, whereas an adult dispenses with its skin four times a year. This molt is a rather comical procedure, at least from the human standpoint. First the skin splits. Then the toad uses its front feet to pull the old skin off, gradually stuffs the castoff suit into its mouth, and finally swallows the whole affair! The entire procedure takes only about five minutes, and the toad then goes blithely on its way.

Our common toad is notably uncritical about terrestrial habitat, which may be one reason for its evolutionary success. Fields, all sorts of forests, moist areas, gardens, yards, and almost anything else are suitable for the toad, as long as it's not excessively hot and dry, with no shade or moist dirt in which to shelter.

Toads are equally unfussy eaters. A wide variety of insects, slugs, earthworms, and assorted other small creepy-crawlies appeal to their catholic tastes and are snapped up with gusto, to the great benefit of many plants—including those in our gardens. Like frogs, toads have the unique feature of a long, sticky tongue attached in the front of the mouth, rather than at the rear. This marvelous adaptation enables the toad to flick its tongue forward with lightning rapidity to snare its unwary prey and just as quickly flip it back into the wide, gaping mouth.

Although predators wreak the greatest havoc on toads while they're either tadpoles or very tiny toadlets first emerging onto land, adult toads aren't immune, but the toxins produced by the glands on a toad's skin repel many predators, especially those that have had previous experience with toads. Domestic dogs and house cats, for instance, are known to learn that toads are distasteful. No doubt many wild predators are equally deterred by the toad's defenses, although raccoons and skunks evidently are predators of toads. Snakes are probably the toad's most serious predators, however. Many species of snakes seem to be unaffected by the toad's toxins, and some snakes feed very heavily on toads.

Toads also have a rather peculiar reaction to danger. When startled or frightened, they may release a quantity of liquid that is generally thought to be urine, but in fact is water stored away by the toad against a time when drought might threaten it. How successful this technique is in repelling predators is questionable. I well recall that, when I was a small boy, my playmates and I used to call toads that reacted in this fashion "pee toads" and giggled a little self-consciously about it, as small children are wont to do about such things.

Like other amphibians, toads are cold-blooded, so with the onset of cold weather they begin to look for a place to hibernate. Suitable locations include burrows in soft soil or the soft, loose litter of the forest floor. In places such as this, toads can become dormant for the winter, secure from bitter weather and the attacks of predators.

Toads become sexually mature when they're three years old. When they emerge from their winter's sleep during their third spring, that deep, ancient instinct to mate and perpetuate the species finally asserts itself. Then the breeding grounds issue their mysterious, irresistible siren call, and the toads temporarily forsake their terrestrial habitat to wend their way in slow toad fashion back to the most ancient habitat of all—water. There they will continue the unending cycle that has repeated itself for countless millions of years.

Despite their newfound favor in some circles, toads have generally been regarded, at best, as rather plebeian creatures—useful, perhaps, but without the slightest hint of glamour or excitement. How ironic it is, therefore, to learn that the lowly toad may yet prove to be the savior of some of its more charismatic relatives.

The Center for Research of Endangered Wildlife (CREW) and the Herpetology Department of the Cincinnati Zoo are busily engaged in developing assisted reproductive techniques to help propagate endangered species of toads and frogs. Oddly enough, their quest started with an old test originally formulated for humans, rather than for frogs and toads.

The Galli-Mainini test was developed fifty years ago to determine human pregnancy. It was discovered that male toads injected with urine from a pregnant woman began to release their sperm, while the urine of non-pregnant women had no effect. Evidently the human pregnancy hormone replicates the hormone normally released in the male toad during amplexus.

The CREW team first tested different methods of introducing human urine into the toad in order to attain maximum effectiveness. Application on the skin of the back had no effect, and application on the skin of the abdomen only induced sperm release in about one-third of the males. Injection under the abdominal skin caused some sperm release, but by far the best results were obtained by injecting the urine into the toad's abdominal cavity.

Using this technique developed with the common toad, CREW and the Cincinnati Zoo have successfully bred endangered toads (and ultimately will breed frogs, one would assume) and released their offspring into the wild. The poor relation, it turns out, is not so poor after all!

During cool, damp weather, toads are active both day and night. When the weather turns hot and dry, however, they burrow into soft soil, forest litter, or beneath some other shelter. There they wait through the heat of the day, emerging at dusk to forage during the cool, damp night.

On numerous hot summer evenings, as my wife and I have sat in front of our house to watch the evening bat circus, we've heard tiny noises as dusk settles—perhaps a slight scratching or scraping, or the movement of a small pebble. Peering around us, we've been able to see a toad that has just emerged from beneath the huge, flat stone that constitutes our doorstep. There, in relative coolness, the toad waited out the heat of the day.

At first the toad's movements are cautious and tentative—a single hop, followed by total immobility for several minutes. Then comes another hop, followed by another pause. Then, as twilight deepens, the toad seems to gather confidence; soon it hops slowly but steadily away, quickly disappearing in the gathering gloom. Toads are abroad again, seeking sustenance, and we're left with the comforting feeling that one more thing is right with the world.

Great horned owl *(rear);* barred owl *(center);* saw-whet owl

11

Not Wise, But Otherwise: Owls

Myths

- Owls are wise.
- Owls only make hooting sounds.
- Owls are blind, or nearly so, in bright daylight.
- An owl's ears are visible.
- Some owls can carry off many times their own weight.

OWLS HAVE BECOME THE UBIQUITOUS SYMBOL OF WISDOM IN MODERN SOCIETY. No matter which way we turn, anything connected with wisdom or intelligence is apt to include a representation of an owl. Advertisements of the "Be wise, modernize" sort, for example, usually sport a drawing or photograph of an owl. Then there is the oft-repeated bit of doggerel, "A wise old owl sat in an oak. The more he saw, the less he spoke . . . Why can't we be like this wise bird?" slightly modified from its original 1875 incarnation in *Punch.*

It was not always thus, however. In the past, owls in many cultures had a sinister reputation. This was quite possibly because owl hoots have generally been considered "mournful," and hence have associated owls with death and a variety of ill fortunes.

Shakespeare certainly believed that, or at least used the common superstition for dramatic effect. In *Macbeth,* Lady Macbeth takes the cry of the owl as a sign that her husband has murdered Duncan ("It was the owl that shriek'd, the fatal bellman . . ."). A little later in the play, the owl becomes a symbol of the weird atmosphere surrounding Duncan's murder: "'Tis unnatural . . . on Tuesday last, a falcon, towering in her pride of place, / Was by a mousing owl hawk'd at and kill'd."

Not everyone considered owls to be the bearers of misfortune, however. For instance, a small owl in ancient Greece was so highly regarded that it became the mascot of the powerful Athenian army. The owl was one of the symbols of the goddess Athena and of the city of Athens. It was portrayed on Athenian coins, with the head of Athena on the reverse side. Other societies also valued the owl as a bearer of good fortune—perhaps even as a symbol of wisdom. It's doubtful, however, that any previous society has placed as much emphasis as ours on the owl's reputed wisdom.

Why have we endowed these birds with such an aura of great wisdom? The answer seems to lie in the owl's facial structure and large, staring eyes. An owl's eyes are set squarely in the front of its head, an avian feature shared only by penguins. This eye placement conspires with the feathered facial disk to give the owl a vaguely human appearance. In addition, those big, staring eyes lend an almost professorial appearance to owls, much as horn-rimmed glasses might to a human. Anthropomorphism does the rest; if owls remind us of ourselves, they must therefore be intelligent.

Appearances notwithstanding, the sad truth is that owls don't rate very high on the intelligence scale, even among birds. Members of the Corvidae—ravens, crows, jays, magpies, and the like—to use just one example, are undoubtedly far smarter than owls. Upon careful consideration, it shouldn't come as any great surprise that owls rate as something of dim bulbs among their bird brethren. After all, so much of an owl's skull capacity is taken up with the apparatus necessary for vision and hearing that there's little brainpower left for "wisdom"!

We actually do owls an injustice by regarding them as wise, for in so doing we obscure their true nature. Owls are, above all else, superb killing machines, with astonishing adaptations for taking their prey. With their almost unbelievably sensitive hearing and eyesight, as well as virtually silent flight, owls possess a phenomenal ability to locate and seize a variety of prey by night or day.

Take eyesight, for instance. An owl's eyes are one hundred times more sensitive to light than human eyes. Even on a dark night, owls can see details of landscape and prey with great clarity. Because most owls are night hunters, it's popularly supposed that in bright daylight they see poorly at best, and are perhaps nearly blind. This is a misunderstanding that dates far back in time. Tennyson, among others, certainly believed this myth when he penned the line ". . . and thrice as blind as any noon-tide owl. . . ." However, owls can see equally well in darkness or daylight.

Why, then, are most owls mainly abroad at night? There are at least two good reasons. First, this is the time when the majority of their prey species are most active; mice and other small rodents, for example, are far more apt to be afoot at night than in daylight. Second, they have no nighttime competition from other raptors—notably hawks and eagles. Equipped by nature with superb tools for night hunting, owls have thus been able to preempt an important ecological niche.

Because of their size, structure, and placement, an owl's eyes are immobile, but the owl compensates with a unique adaptation. In both birds and mammals, the first vertebra below the skull is called the *atlas* because, as Atlas supported the globe, it holds up the skull. Mammals have an atlas with two cups that fit against two projections on the skull; this arrangement severely limits the ability to turn the head.

Birds, on the other hand, have a single cup-and-skull projection, a system that greatly facilitates movement of the head. Owls have further perfected this useful anatomical feature. In consequence, an owl can swivel its head not merely 180 degrees to look backward, but 270 degrees—three-quarters of a full circle.

I first observed this trait many years ago while watching Spooky, the resident great horned owl *(Bubo virginianus)* at the Museum of Science in Boston. Fascinated by its huge, yellow eyes, I somehow conceived the idea of a staring contest to see who would blink first. It was a difficult struggle at first, for I had to exercise a great deal of determination to avoid blinking. After a substantial time, however, I "won" the contest—not because Spooky blinked, but because I suddenly found myself staring at the back of its head!

After a few seconds the owl snapped its head forward again, and a somewhat briefer staring contest ensued. Then I once more found myself viewing the back of Spooky's head. This sequence was repeated several times more, with each staring episode becoming shorter and shorter. Finally it reached the point where the owl was taking only a split second to see that I was still staring—then snapping its head backward, then forward, then backward, and so on. At that point, fearful that the owl would twist its head off, I took pity on the distraught creature and left poor Spooky in peace. Clearly, a great horned owl doesn't appreciate having its stare returned—at least for any length of time!

Owls' hearing is even more astonishing than their eyesight. Indeed, despite superb nocturnal vision, owls hunt at night more by sound than sight. So incredibly sensitive is their hearing that, according to researcher Allen Eckert, some species of owls can hear a mouse squeak as much as a half-mile away! An owl's hearing ability consists of far more than mere sensitivity, however; it also has a special adaptation enabling it to locate its prey's position in three dimensions with uncanny precision. Indeed, experiments in the laboratory have demonstrated that sightless owls can catch mice by sound alone.

The ear cavities in an owl's skull are asymmetrical, so that one ear receives more sound from above, the other more from below. This ear placement permits the owl to locate its prey accurately in the vertical plane. Simultaneously, an acute ability to process the infinitesimal difference in the time it takes a sound to reach each ear accomplishes the same thing in the horizontal plane. Also, the facial disk—a feature that helps give owls a vaguely human appearance—further aids this phenomenal hearing apparatus by capturing sound and reflecting it toward the ears.

If the prey makes the slightest sound while an owl homes in on it, the owl immediately "views" the location of the prey in three dimensions. Incidentally, what are widely regarded as "ears" on some owls are merely feather tufts that have nothing to do with hearing. The actual ears are only openings in the skull and are invisible.

Silent flight completes the package of adaptations that make the owl such a ruthlessly efficient predator. Exceptionally soft leading edges on flight feathers, combined with a very large wing area in relation to body weight, plus soft head and body feathers, make an owl's flight almost noiseless. The mouse rustling in the grass or the rabbit hopping across a forest glade has no warning that death is descending upon it as the owl ghosts down for the kill. Not until talons seize the prey in a deadly grip does the victim have the slightest inkling of its fate. These days we make much of the technological wizardry of our stealth fighters and stealth bombers, but owls on ghostlike wings were nature's original stealth aircraft long ages earlier.

Owls, even large species, subsist mostly on small mammals such as mice, voles, and lemmings. These are promptly gulped down whole, and then another adaptation—their unusual digestive system—takes over. Indigestible parts, such as bones and fur, are packed together into pellets by stomach action, then are regurgitated, thus efficiently ridding the owl of useless material. An owl's favorite roosting place can sometimes be identified by the pellets beneath it. These pellets can also tell us much about an owl's food habits. By dissecting and analyzing pellets, researchers can determine what species an owl has been feeding on.

Collectively, owls are properly known for their hooting, but most people fail to realize that they make a wide variety of other sounds—and that some species don't hoot at all. Many strange nighttime noises, and some during daylight hours as well, are often attributed to various other creatures, when in fact they emanate from the vocal apparatus of owls. Yes, some owls do give a hoot, but this is by no means their universal language.

At least eighteen species of owls are found in the United States and Canada. Several of these, however, such as the elf owl and ferruginous owl, are found mainly from the southern part of Mexico and barely come into the United States. It's beyond the scope of this book to attempt a description of every species and its habits, but let's take a close look at several of the more widely distributed and interesting owls.

The Barred Owl

The barred owl *(Strix varia)* is the most common large owl in the eastern United States, ranging to roughly the Mississippi River and considerably farther west in southern Canada. Its range also extends northward to the southern tip of Hudson Bay. This is also the most commonly seen owl because it's often quite active in the daytime, especially in the late afternoon. Named for the horizontal bars across the upper breast and the prominent vertical brown streaks down its belly, this owl is easy to identify because of its big, liquid brown eyes and lack of ear tufts. (The barn owl is also "earless," with brown eyes, but is much smaller and has a heart-shaped "monkey" face.) In contrast to the fierce yellow eyes of most owls, the barred owl's big brown eyes lend it a deceptively benevolent appearance.

With its puffy head and a wingspan of three and a half feet or more, the barred owl is an imposing bird. Among owls commonly resident in the United States, only the great horned owl is larger; in Canada, the great gray and snowy owls are larger, as well. Still, the barred owl's size isn't far inferior to that of those other species, and anyone who has had a close view of one would certainly describe it as a large owl.

Despite its rather impressive height and wingspan, the barred owl can serve as Exhibit A in documenting why owls in general aren't what they appear to be. Beneath all those feathers, there's remarkably little bird: even a large barred owl weighs only about two pounds, and most are probably closer to a pound and a half or less! There's a purpose to all those fluffy feathers, though: they serve the dual function of helping the owl flit soundlessly through the air and insulating it against the bitter winds of winter.

Mice and voles (so-called meadow mice) constitute the main part of the barred owl's diet. It also preys on frogs, large insects, and occasionally squirrels and small birds. Although barred owls might occasionally seize prey as large as a rabbit, small creatures are by far their favored food.

In normal times, at my home we see barred owls only occasionally, mostly while out in the woods. Whenever the snow becomes very deep, however, we're apt to have a barred owl frequent the vicinity of our bird feeders, often in the daytime. With truly astonishing originality we've christened this visitor, who has taught us much about the barred's preference in food, Shakespeare. Actually, we have no idea whether we've been seeing one Shakespeare or several Shakespeares, since it's rather difficult to tell one barred owl from another, especially from year to year.

As previously noted, our bird feeders are utilized by many red squirrels in the winter. Barred owls are known to kill squirrels, and a naturalist friend once reported seeing a barred owl make off with a red squirrel in the woods. Yet in all the time we've watched Shakespeare, he/she has only made one serious attempt and another rather halfhearted one at a red squirrel. During the same period, not a single attack has been made on any of the many birds regularly using our feeders.

The one serious attack on a red squirrel came after I had been watching Shakespeare for at least a half hour. The owl was plainly visible, sitting on a low branch no more than twenty yards from the feeder, yet neither birds nor squirrels paid it the slightest attention.

One rather reckless red squirrel, in particular, seemed to be tempting fate by running back and forth in the direction of the owl, seeking sunflower seeds that birds or wind had dispersed hither and yon. Often the owl swiveled its head to gaze at the squirrel, yet made no move to attack. At last, though, as if goaded beyond endurance by the audacity of this particular squirrel, Shakespeare launched himself without warning into a long, swooping dive.

Alas for the owl's plans, Big Red wasn't quite as foolish as he appeared to be. We had stuck our Christmas tree in the snow between the two feeders; the squirrel raced madly toward it and, just as Shakespeare seemed about to strike, leaped frantically into the protective branches, just inches ahead of the owl's reaching talons.

Shakespeare's other attempt at a red squirrel was far less dramatic. Again, it occurred after what might be considered extreme provocation, but this attack seemed leisurely, merely *pro forma,* as if to say, "I'm here, and don't forget it!"

No doubt a barred owl's decision to try for birds and red squirrels depends a great deal on how famished it is. After all, any predator, if sufficiently hungry, will go after prey that may in better times be well down its preferred list of species.

If such things as birds and squirrels aren't the most tempting of morsels to a barred owl, mice are in a different league entirely. Based on two personal experiences, I would say that owls regard mice and voles in about the same light as we might view lobster, steak, or a rich chocolate dessert. In fact, mice were the vehicles for my two most memorable encounters with a barred owl.

We keep a container of sunflower seeds in our basement garage, and two deer mice took up residence in it. The container is made of heavy cardboard, and the mice first gnawed a hole through the top, then scraped shavings off the inside to begin making a nest. I like mice outdoors; they're cute little beasts, and fun to watch. They don't belong inside the house, however, where they're destructive and dirty, and where they multiply at an alarming rate. Still, I was reluctant to kill the little creatures without trying to get them out of the house first.

For three or four days I took the seed container outside each day, tipped it on its side, dislodged the mice, and hoped they'd take the hint. Fat chance! The next day they were always back in the container, expanding their nest in preparation, one would assume, for a new crop of mice. The situation was intolerable, and my patience finally ran out.

Then Shakespeare appeared for one of his periodic but unpredictable visits. At that point the light went on: Shakespeare plus mice equaled problem solved. Further, I thought this might be a golden opportunity for some great owl photographs. Accordingly, I slung a camera around my neck, picked up container and mice, and went outdoors.

Shakespeare was perched on a low branch about fifteen feet away, perhaps a dozen feet above the ground. I put the container on its side, took my camera in one hand, and slapped the container with the other to scare out the mice. Just as the first mouse started out of the container, I lifted my eyes to view Shakespeare's reactions and was stunned to see that the owl had already launched and was down to eye level, coming straight at me!

Before I could even react, the owl landed in the snow with a *flop* no more than two feet from me. There, for two or three long seconds, we stared at each other, while I looked deeply into those liquid brown eyes. Believe me, at that range there was not the slightest sign of the benevolent appearance that distance had always lent. Instead, there was a wild fierceness in that gaze that coursed through me like an electric shock and made me everlastingly grateful that I was so much larger than the owl. Then the spell was broken, and the owl levitated effortlessly to its former perch, *sans,* unfortunately, either rodent.

The sequel to this extraordinary encounter took place the following day. The mice were back in the container, and this time Shakespeare was ensconced on a slightly more distant perch, perhaps sixty or seventy feet away. Again I booted the mice out of the container, this time a bit better prepared for the owl's reaction. As before, Shakespeare launched instantaneously when the mice appeared on the run, headed in his direction. Down he swooped, as swiftly as an avenging Fury, plucked up one of the mice, and made off with it into the woods. The other mouse, by the way, never returned to the feed container.

The barred owl is not only our most commonly seen owl, but also by far our most vocal owl, both in the quantity and great variety of its calls. Its colloquial name of "eight hooter" stems from its most typical call, the familiar *hoohoohoohooo, hoohoohoohoooaww.* The *aww* at the end frequently isn't audible at any distance; close up, it's a raspy, almost gargled sound on a slightly lower note than the rest of the call.

The basic eight-hoot call of the barred owl is quite easy to imitate with just a little practice, and owls can often be induced to respond, or even be called closer, in this fashion. It's also possible to purchase calls that can be blown to give a passable imitation of a barred owl.

There are many variations in the number and pattern of hoots, ranging from a single *hoooaww* to a four-note *hoohoohoohooooaww,* along with all sorts of in-between permutations. We've often used the technique of trying to imitate precisely whatever a barred owl says, especially after we've called one to close quarters. On a number of camping trips, we've carried on lengthy conversations with barred owls—sometimes two owls at a time—by imitating their every call. Sometimes these conversations have stretched out for nearly an hour until either we or the owls grew tired of it. I have no idea what we were saying in owl language on those occasions, but it must have been good!

This wide variety of hoots by no means exhausts the repertoire of such a remarkable vocalist. In addition to hoots, this bird can emit a bloodcurdling scream that is truly frightening to the uninitiated. Some have described it as sounding like a woman's scream. Others, unfamiliar with the barred owl's vocal versatility, have attributed it to bobcats and other beasts. After all, owls are supposed to hoot, not shriek like a banshee. Once this scream has been heard a few times, its source can immediately be identified as a barred owl:

instead of starting as a full-throated scream, such as a human would make, it begins as a thinner sound that builds steadily to a crescendo, thus:

eeeeeeeeeeeeeeeeeeeeeeaaaaaa.

This cry becomes amusing, rather than frightening, when one becomes familiar with it.

Far more amusing, however, is what I term the "monkey call." This rather staccato production, *oohoohahhahhawhawhawhawhaw,* is a bit difficult to portray in print, but it indeed sounds almost like the call of certain kinds of monkeys. Hearing a pair of barred owls close together, with one initiating this call a note or two after the other, is sure to elicit a delighted laugh from the human listener!

In addition to these calls, as well as hisses, moans, and all sorts of other variations, fledgling barred owls make yet another sound. Early one summer we became aware of frequent little mewing noises, much like a cat's but softer, coming from the woods close to our house. My wife suggested that, since most of the strange noises emanating from our woods were made by barred owls, this was probably an owl call, as well.

I thought this was rather dubious, but finally took binoculars and sneaked very slowly and quietly toward the source of the mewing. Before long I spied a movement in a nearby tree, and the binoculars revealed a pair of fledgling barred owls, not long out of the nest and still a bit fuzzy. Each time their beaks opened, I could hear the soft mewing sounds. Probably the pair were importuning their parents for food, though there was no sign that the adults responded to these piteous little cries.

An added feature of the barred owl's vocalism is that it often calls in the daytime, especially from midafternoon on. Why barred owls spend so much time calling at all hours and all seasons of the year is something of a mystery. Perhaps these calls are territorial at times, mating calls at others, and perhaps merely conversational most of the time.

Barred owls nest relatively early in the spring, usually from late March to the end of April. A cavity in a tree is the preferred nest site, although an old crow's or hawk's nest may be appropriated if tree cavities are scarce. Usually two or three eggs are laid—though occasionally as many as four. The eggs hatch in about a month, and it takes the nestlings roughly another month to fledge. Then they disperse to seek their own territories and call back and forth as the spirit moves them.

The Great Horned Owl

If the barred owl's brown eyes give a false impression of mildness, the same can't be said for the great horned owl. With its large, staring yellow eyes, long ear tufts, and great size, this owl exudes fierceness from every feather. Furthermore, the great horned owl is every bit as fearsome a predator as its appearance implies. The largest "eared" owl in North America, a large specimen can stand two feet high and have a wingspan of up to five feet. As with other owls, though, most of this bulk is feathers. An exceptionally large great horned owl may exceed five pounds, but three and a half pounds is about average—not much heft for a bird with such an impressive wingspread. But despite this rather puny body, the horned owl is all muscle and sinew.

Where most other owls content themselves mainly with mice and other small creatures, the great horned regularly seizes rabbits and hares, grouse, crows, weasels, opossums, muskrats, and other prey of similar size. Many a wandering house cat has been snatched by those vicious talons, and the great horned is probably *the* major predator of skunks: evidently lacking a sense of smell, like many birds, the owl seems unfazed by the skunk's potent olfactory assaults.

I became aware rather early on of the horned owl's penchant for dining on skunk. At about the age of ten, a friend and I took a correspondence course in taxidermy. We weren't very good at it, but our efforts with birds came out reasonably well—though only because the feathers covered up a multitude of errors!

In those days, people shot large owls on sight—a practice that was at least occasionally justified, as readers will learn. In any event, someone brought me an exceptionally large great horned owl to be mounted, and the bird reeked of skunk scent. I held my nose, stuffed the owl, and absorbed the information that a great horned owl was willing to tackle rather large prey, including skunks.

There were other things that I knew about great horned owls, as well. I was an extraordinarily fortunate child in many ways. My father and mother lived on my maternal grandparents' farm while I was growing up. There my parents read to me extensively, while my grandmother, to my great delight, told me a wide variety of stories. Like every child, I loved to hear the same stories over and over, and my absolute favorite was the true story that I called "It Looks Like the Work of an Owl."

We always had hens on the farm, kept in a hen house at night but roaming freely from dawn to dark. Many years before, hens began to mysteriously disappear. Then one day my grandfather came across the partly eaten carcass of the most recent loss. He examined it carefully and then reported to my grandmother, "Well, Luella (a nickname for Lucy Ellen), it looks like the work of an owl!"

Grandpa set a trap beside the carcass that night, and when he went to milk the cows the next morning, there was the owl in the trap. Grandpa was in a hurry to start the morning chores, so he decided to kill the owl when chores were done. When he returned, however, the owl had departed, leaving one claw in the trap.

Undeterred, Grandpa reset the trap, and the owl—displaying a noteworthy lack of wisdom—was once again in the trap the following morning. This time, Grandpa wasted no time in dispatching the owl! There were no more strange disappearances among the flock of hens thereafter.

Although flocks of free-range hens are relatively uncommon these days, marauding great horned owls can still cause trouble from time to time. A friend who is a cabinetmaker told me about just such an incident that took place not many years ago. His shop was connected to the chicken coop, and one night, while working on some furniture, a great uproar erupted among the hens.

My friend ran outside to the coop's entrance, and by the light coming into the coop from his shop, he saw a huge owl, wings outspread, clutching a hen. He was afraid to enter the coop for fear the owl might fly into his face, so he poked a broom inside and vigorously swished it around. This disruption soon became too much for the owl, which loosed its hold on the hen and departed in an indignant flurry of wings. The hen recovered, according to my friend, but the attack must have severely traumatized her, for she never laid another egg!

The great horned is the most widely distributed North American owl, inhabiting the continent coast to coast from the Arctic in Alaska and part of Canada far down into Mexico. Despite its wide range, however, the great horned owl is far less common than the barred owl. It's also seldom seen, not because it's rare, but because it's much more nocturnal than the barred owl.

If the great horned owl is less visible than its barred brethren, it's also far less vocal, both in quantity and variety of calls. Its most common call is a series of deep, resonant hoots, almost always heard at night. It can also utter a series of low hoots that sound almost like the cooing of a dove, although deeper and throatier. Like most owls, the great horned will also hiss when alarmed or upset. Owing to its nocturnal habits and infrequent calls, the presence of this big owl often goes undetected.

The great horned owl fully deserves its reputation as a fierce predator, but this attribute is sometimes exaggerated. For example, one male great horned owl caused all sorts of problems, even attacking humans—quite possibly in defense of a nearby nest. According to various accounts, the owl attacked a twenty-pound dog and flew away with it. The dog's owner then ran after it, shouting at the owl, which dropped the dog from a height of thirty feet. The dog died; the owl, when later killed, weighed three and a half pounds.

This feat struck me as somewhere between extraordinarily improbable and just plain impossible. Leading ornithologists share my skepticism. As Dr. Stuart Houston, one of the foremost great horned owl experts in North America, put it, "No bird can carry six times its weight." He and other ornithologists believe that a great horned owl, with its fearsome talons, could kill a twenty-pound animal, but carrying it off is another matter entirely.

A three-and-a-half-pound great horned owl might possibly be able to carry as much as five or six pounds for a short distance—a rather amazing feat for a bird—but certainly not twenty pounds. There was general agreement among the ornithologists with whom I consulted that either the dog weighed a great deal less than twenty pounds or the owl killed it but didn't carry it off. The great horned owl is an amazing bird, but it doesn't possess supernatural powers!

The great horned owl has the distinction of being perhaps the earliest nester of any bird throughout much of its range. Nesting may begin as early as January, with snow piled deep on the ground, and it's not uncommon for the female, superbly insulated by her thick coat of fluffy feathers, to incubate her eggs while covered with a mantle of snow.

Great horned owls possess many skills, but nest building isn't among them. In forested areas, horned owls most commonly lay claim to an old hawk's, crow's, or heron's nest, but a large tree cavity sometimes serves the purpose equally well. In areas that lack suitable tree sites, the female may simply lay her eggs on a ledge.

Once a nest is established, the female lays one to three eggs, although the usual number is two. Incubation takes about a month. Thereafter, the harried parents are forced to hustle after enough meat to satisfy the rapidly growing appetites of their voracious youngsters. The gawky, homely, fuzzy young grow rapidly and, after about a month and a half, have become sufficiently feathered to fledge and leave the nest. If owls can feel anything like relief, their parents must surely experience it at this point!

THE SNOWY OWL

This Arctic resident is a birdwatcher's delight when it descends into southern Canada, the northern tier of states, and, very occasionally, as far south as Florida, Texas, and central California. Why? There are several reasons. First, this is a big, showy owl. With a wingspan of nearly five feet and a weight of four to six pounds, it's the heaviest of our North American owls. It usually stands out like a beacon when it moves south, because of its imposing size and spectacular plumage—nearly snow white in adult males, more speckled in adult females, and barred with black in juveniles. Further, the snowy owl *(Nyctea scandiaca),* while by no means a rare winter visitor far south of its tundra breeding grounds, is just uncommon enough in many areas to create a stir. Added to all this are the snowy owl's unusual eyes; whereas most other owls have round eyes, the snowy owl's yellow eyes are narrowed somewhat, like a cat's.

Finally, snowy owls have the endearing trait—at least to birdwatchers—of staying in one place for hours on end, usually prominently displayed on a dead stub, a utility pole, or even the top of a building. This is no mere accident: snowy owls evolved as hunters on the vast, barren Arctic tundra, where they prefer to perch on the highest point around and wait until they spot their prey—then glide down to seize their victims by stealth. Thus, when they visit southern Canada and the United States, the big predators favor wide-open spaces (airfields such as Boston's Logan Airport are often preferred hangouts) and high perches, where they can approximate tundra hunting conditions.

Snowy owls do much of their hunting diurnally. This is no great surprise, considering that there is daylight almost twenty-four hours a day during their high Arctic breeding and nesting season. Conversely, they must also be efficient night hunters during the long stretches of almost total Arctic darkness.

Summer prey for snowy owls consists almost entirely of mammals—mostly small, with lemmings making up the bulk of their diet. In winter, especially for those owls that migrate south, their meals are far more varied. Hares and ptarmigan help carry the owls through the winter in the Arctic, when lemmings are mostly active beneath the snow. Owls wintering farther south have proved quite adaptable when it comes to prey. Mice are a staple, but Norway rats are also prime fare. For that matter, so are pigeons, rabbits, dead fish, and almost anything else of suitable size that comes to the owls' attention.

It was once thought that these white visitors from the Arctic came south in winter because of a shortage of lemmings. Although it's a complete myth

Snowy owl

that lemmings periodically commit suicide by throwing themselves off cliffs into the sea, where they drown en masse, the plump little rodents are notoriously cyclical, going from almost unbelievably high populations to extreme scarcity every four or five years. Unquestionably, lemming numbers have an effect on snowy owl populations, but biologists are learning that the interrelationship between these two species is far more complex than has heretofore been suspected.

For one thing, there's no evidence that lemming cycles are synchronized throughout the Arctic, and they may be quite regional. Since snowy owls by nature are great travelers, it's no special feat for them to move from an area of lemming scarcity to one of abundance. For another, large numbers of snowy owls migrate annually to the Great Plains area of Canada and the United States without apparent reference to lemming cycles. Much remains to be learned about the dynamics of the lemming/snowy owl relationship.

Snowy owls are silent for most of the year. During the breeding season, however, they utter a variety of sounds, especially a croaking call and a sort of shrill whistle. They may hiss when they or their nests are threatened, and reputedly also hoot during the breeding season.

Prior to breeding, snowy owls perform a fascinating courtship ritual. First comes the flight display, in which the male alternately descends with wings arched above his back, then flaps upward, only to repeat the procedure. Then he begins to bring lemmings, which he deposits in a pile in front of the female; this performance may signal that he's a good provider, that prey is sufficiently abundant to enable the parents to raise a brood successfully, or both. Finally the male goes through a variety of strange poses and then fawns in front of the female to complete his amorous display.

Like the rest of their tribe, snowy owls aren't nest builders. On a hummock or other high point, the female simply scrapes out an unlined hollow and begins to lay eggs. It's noteworthy that such a location provides not the slightest protection from the bitter Arctic winds, but allows the nesting pair to watch for predators in every direction. Protection from the wind evidently isn't important to snowy owls, for their magnificent plumage has the same insulating power as that of Antarctic penguins and enables them to withstand the horrific cold of an Arctic winter.

At this point the species' nesting behavior becomes highly unusual. Snowy owl clutches range from the merely very large—for owls—of five to eight eggs to the huge, with as many as sixteen. The exceptionally large clutches are thought to occur only in years when lemmings are extremely abundant, but that's by no means certain. What is certain is that the female lays one egg roughly every two days, yet she must begin incubation as soon as the first egg is laid, lest it freeze solid in the frigid Arctic weather. If she lays a dozen eggs, this means that nearly a month elapses between laying the first and last eggs in the clutch. Each egg takes about thirty-two days to hatch, so the last egg has barely been laid by the time the first one hatches!

With the nest on the ground, the fuzzy gray owlets don't have to fledge in order to leave it. They merely step out of the nest, long before they can fly, and nestlings from the early eggs walk away soon after siblings from the later eggs have hatched. The owlets move about, but remain near the nest, where the male brings them food. The female is tied to the nest until the last owlet leaves, which can be two months in the case of a very large clutch.

Snowy owlets fly about seven weeks after hatching, so raising an exceptionally large brood, from the time the first egg is laid until the last owlet takes wing, can consume more than three and a half months. Since that represents virtually the entire Arctic summer, the process of raising such a large brood places an enormous strain on the adult pair, particularly the male, who must do all the hunting while the female is laying and incubating the eggs. Adult snowy owls consume as much as four or five lemmings a day, and the appetites of their young grow daily, so it's been estimated that a pair of adults with a brood of, say, eight or nine can devour more than 2,500 lemmings in a single Arctic summer!

The Saw-whet Owl

We tend to think of owls as large birds, but they come in all sizes, from the very large to the middle-sized to the tiny—and one of the tiniest is the saw-whet owl *(Aegolius acadicus)*. How small is tiny? The diminutive saw-whet stands only seven or eight inches tall and weighs approximately three ounces. It can easily fit in the palm of one's hand and somehow seems too small to be a raptor like its bigger relatives.

Where other owls appear benign, comical, or fierce, the saw-whet, despite yellow eyes, can only manage to look cute—an owlish version of Tom Thumb. This mien is perhaps reinforced by the fact that the saw-whet is "earless," with a smooth, rounded head. There is nothing cute about this diminutive owl when it comes to predation, however. Make no mistake, those little talons are strong and needle-sharp, and the tiny beak is perfectly capable of dissecting prey with exemplary efficiency.

Although the saw-whet eats insects and, occasionally, small birds—prey that would seem in keeping with its size—its main diet consists of mice, voles, shrews, and even young squirrels. Consider that a fat meadow vole approximates the weight of the saw-whet itself, and a young squirrel substantially outweighs it, and the rapacious nature of this little owl comes into clear focus.

The saw-whet is primarily a northern owl—resident across southern Canada and southern Alaska, the northern portions of the United States, and south through the Rocky Mountains and along the Pacific Coast to Mexico. Although it doesn't migrate in the true sense of the word, the population of saw-whets does shift somewhat southward during the winter. In the eastern United States, the saw-whet often winters as far south as the Carolinas; less commonly, it may even be found in winter as far south as the Gulf Coast.

Although this tiny owl is common and wide-ranging, it's rarely seen for two reasons. First, the saw-whet is almost entirely nocturnal and spends its days roosting mostly in dense evergreens, where its small size and brown coloring render it nearly invisible. Perhaps that's why I've never been privileged to see a live saw-whet. Incidentally, saw-whets are noted for their great tameness when roosting during the day. They seem totally unconcerned by human proximity and reputedly can even be handled at such times.

In partial recompense for never having seen a live saw-whet, I've been fortunate enough to have two dead ones brought to me. My first sight of a saw-whet came only a couple of years after I mounted the skunk-scented great horned owl. A lady had found the little thing in her barn, evidently dead from natural causes. I mounted it at her request, constantly amazed that an owl could be so tiny. A second saw-whet was given to me much more recently; it, too, had died of natural causes—very possibly starvation, since it was found in the middle of a severe winter.

The second reason why the saw-whet is so seldom seen is that it's silent for most of the year. Only with the approach of spring does the little owl vocalize, emitting a sustained series of rapid whistles, most frequently described as a monotonous *tootootootootoo,* well over one hundred times a minute. Clearly, the saw-whet is the Johnny One-Note of the bird world.

To the best of my knowledge, I've never heard a saw-whet owl, but an acquaintance told me about her initiation with its "song." For hours on end, while she worked around the house on an early spring day, she kept hearing a strange and unfamiliar noise. She described it as sounding like a radio test signal—a sustained series of very short, high-pitched beeps—which went on and on until they nearly drove her crazy.

The unfortunate woman began to fear that there was either something wrong with her hearing or that she was losing her mind. Finally, in near desperation, she called in someone else and asked if that person could hear the noise. When the answer was affirmative, she was first greatly relieved and then set about locating the source of the sound. After some effort, and a bit of aural triangulation, she finally found it—a little saw-whet perched demurely in an evergreen tree behind her house.

Like other owls, the saw-whet eschews any nest-building chores. Nesting takes place in a tree cavity, often in holes abandoned by woodpeckers; abandoned flicker nests appear to be a favorite. Nesting can start as early as the end of March, but may be a month or two later, with five or six eggs laid on the floor of the cavity. Incubation takes three to four weeks before the next generation of this minuscule owl hatches.

Stereotypes being what they are, owls will no doubt continue to symbolize all that is wise for many people, but they're clearly otherwise—far from the brightest of birds. That shouldn't diminish our admiration for them one whit, however. Quite the contrary, owls have such astonishing abilities as predators that we should view them with respect and wonder. With their wide variety of sizes and habits, and their many amazing adaptations, owls of all descriptions will richly repay any time spent in studying or observing them.

12

Which Is Which? The Heron and the Crane

MYTHS

- Herons and cranes are the same thing.
- All North American cranes are endangered.
- Cranes eat mostly fish.
- Both cranes and herons roost and nest in trees.

IN ONE OF THE MOST PREVALENT CASES OF MISTAKEN IDENTITY IN THE REALM OF WILDLIFE, OUR LARGEST HERON, THE GREAT BLUE *(ARDEA HERODIAS)* IS TIME AND AGAIN REFERRED TO AS A CRANE. Over and over I've heard people here in the Northeast, for example, call a great blue heron a crane; alas for accuracy, the only cranes known in this region, historically or otherwise, are fireplace cranes and construction cranes!

Perhaps this misidentification isn't too surprising. To those with only a passing interest in wildlife, there are several similarities between the great blue heron and our North American cranes: both are tall, long-legged, imposing, and most commonly found around water. Although these similarities are somewhat superficial, they can undeniably be misleading. Whatever their apparent resemblance, however, the great blue and the cranes aren't even closely related, and belong to entirely separate families. As will become evident, their differences are much greater than their similarities.

Most people who are unfamiliar with cranes automatically assume they're an endangered species. That, of course, is because the splendid whooping crane *(Grus americana)* is indeed endangered; in fact, it was what might be called the original poster child for North American species of wildlife that are endangered or threatened. The sandhill crane *(Grus canadensis)*, how-

Whooping crane *(rear);* great blue heron

ever, is far from endangered. Some 500,000 migrate annually through Nebraska's Platte River Valley, while another 30,000 to 35,000 migrate from the western Great Lakes region to Florida.

A few similarities in general appearance notwithstanding, physical traits are as good a place as any to begin distinguishing between herons and cranes—in this instance, between the great blue heron and the whooping crane. Coloration is a very obvious difference. The great blue is well named for the bluish gray plumage on its body, while the adult whooper is all white except for red around the face and black primary wing feathers (immature whoopers show some rusty coloring). There is, incidentally, a white color morph of the great blue heron, known as the great white heron, but this is found only from southern Florida southward.

If the whooping crane's vivid red face and the bold contrast of black on white feathers make it one of the most strikingly dramatic of all birds, the great blue offers its own special brand of personal adornment. As the spring mating season draws near, the big herons develop an array of plumes flaring from the lower neck, as well as plumes along the back. At the same time, black crest plumes, starting at the eye, bisect the white face and extend well back beyond the head. Together with black, buff, cream, and reddish brown tones on various portions of the neck, these plumes lend the great blue a rare beauty.

In flight, whoopers are easily distinguished from great blues in two ways. First, a crane flies with its head and neck extended straight forward, while the long neck of the great blue is folded back so that its head is almost between its shoulders. Second, the whooper flies with a quick, distinct emphasis on each upstroke of its wings; in contrast, the great blue flies with a slow, steady, rhythmic wingbeat.

On the ground, the differences in the neck are also highly visible. The whooper's neck curves down from the shoulders and then upward to the head in a simple U, while the great blue's neck loops into an S curve, much as it does in flight.

Then there is the matter of the tail assembly. When at rest, the great blue's wingtips and tail blend together almost seamlessly into a slightly blunt point. The whooper, on the other hand, has long plumes that extend from the upper rear and drape down over tail and wingtips in what is often described as a "bustle" effect.

The voices of the two species are also totally dissimilar. As its name implies, the whooping crane has a high-pitched, rather musical cry. In stark contrast, great blue herons generally manage nothing better than harsh, dismal, and decidedly unmusical croaks. When nesting, they contrive to utter a much wider variety of sounds, but these strange cries could hardly be described as musical, either.

The eating habits of the two species are quite different, too. Herons, although their diet can be quite varied, are preeminently fish catchers. This trait has long been recognized. For example, an English writer in 1579 stated, in charmingly quaint language, "Herones, Bitternes . . . These Fowles be Fishers." Indeed, so dependent are great blues on being able to catch and swallow large fish that the inability of many juveniles to master this art is considered a major reason why only one in five survives to adulthood.

Patience is a virtue widely urged on would-be Izaak Waltons, and it's difficult to conjure up a better mentor in this regard than the great blue. Observing this heron at work varies from something akin to watching a video at its slowest possible speed to gazing intently at a still life for an hour or two. A wading heron moves forward, ever so slowly, barely moving one leg at a time, often standing poised on a single leg while it surveys the water for prey. When standing still, the bird freezes into total immobility, anchored stoically in one spot for improbably long periods of time.

There is nothing lethargic about a great blue when it centers a victim in its sights, however. Then, with astonishing rapidity, the elongated neck uncoils, the long, sharp beak cleaves the water's surface with lightning speed, and more often than not the triumphant heron successfully grasps or spears its meal.

If the prey is small, the heron merely points its bill skyward and gulps it down. A large fish or other sizable prey requires more cautious treatment, though. First the heron must totally vanquish the lively and thoroughly uncooperative prey lately hoisted from its watery domain. In service of that goal, it may repeatedly stab its prospective meal to render it more or less inert. This is by no means the only weapon in the great blue's arsenal. While watching a great blue fishing in his pond, a friend saw the heron dredge out a bullpout—a small catfish also known as a bullhead or horned pout—about a foot long. The heron then waded to the bank and industriously pounded the fish on the hard ground. It repeated this performance several times until the fish expired, and only then did the bird attempt to swallow it.

The great blue's seemingly slender neck has an amazing ability to expand so that large fish and other sizable prey can travel from beak to stomach. Nonetheless, swallowing large fish presents a substantial hazard to the inexperienced heron. If the bird attempts to swallow the fish tailfirst, the spiny fins are likely to catch in the bird's throat, condemning it to a slow and thoroughly unpleasant death. This is only one more of the many lessons the juvenile heron must learn successfully in order to survive.

Although large fish are ultimately the key to their long-term survival, great blue herons indulge in a wide variety of other foods. These include frogs, salamanders, crayfish, and insects, but also encompass more substantial fare, such as the occasional small muskrat or duck. At times, great blues will also seek terrestrial food. Mice, voles, and gophers are typical of this cuisine, but they've also been known to seize and devour Norway rats.

If great blue herons are large consumers of fish, cranes most definitely are not. Although cranes will eat fish if they're readily available, they're by no means patient fishermen in the mode of the great blue heron, nor do fish make up an appreciable part of their diet. Instead, cranes stalk about in marshes or on prairie land, gobbling up an extensive array of foods. Crustaceans, amphibians, and insects make up a good deal of their diet, but whoopers also feed heavily on grain and new shoots of grass. Mice and other small rodents are also fair game for the whooper.

As previously noted, the sandhill crane is thriving; in fact, it's so abundant that it's a highly prized game bird, legally hunted along its fall migration route in the Central Flyway. At this point, the belief that herons and cranes are essentially alike intersects with the notion that all North American cranes are endangered or are like herons. The result is widespread incredulity that sandhill cranes are legal game.

Many people, even when they learn that sandhill cranes are abundant, ask, "But why kill a sandhill? Its flesh must be strong and fishy, so what's the point?" Again, this demonstrates the confusion between cranes and herons. Herons, if legal to hunt, would indeed taste strong and fishy, because of their diet. Migrating sandhill cranes, on the other hand, feed extensively on grain and other vegetation. As a result, they're considered a delicacy. I can personally attest to their edibility; I once had the opportunity to sample sandhill crane at a game dinner and found it similar to wild goose—dark, succulent, and extremely tasty, without the slightest hint of any strong, fishy flavor.

A good four feet tall, with a wingspan of about six feet, blue herons are impressively large birds, highly visible and likely to attract the attention of all who see them. They have the additional virtue of being both common and widespread in much of the United States, as well as some of southern Canada. In fact, there are presently estimated to be about 133,000 nesting pairs of great blue herons in North America, with a total population that probably approaches a half-million.

Even larger than the great blue, the whooping crane towers nearly five feet in height and is the tallest bird in North America. With a wingspan of seven and a half feet, whoopers in flight are truly majestic, one of the most stirring sights in all of our continent's rich wildlife heritage.

Nowhere are the differences between great blue herons and whooping cranes more distinct than in their nesting habits. Great blues nest in colonies, known variously as rookeries or heronries. There they build huge nests out of sticks, sometimes locating them one hundred or more feet above the ground, in the tallest suitable trees available. So closely are nests crowded into a rookery that a single tree may bear a virtual village of the crude but impressive nests! Some heronries are extremely large and may contain as many as two thousand adult great blues, to say nothing of their numerous offspring. However, heron rookeries come in all sizes, and some may harbor a mere handful of nests. In rare cases, great blues may even be solitary nesters.

Scientists are still debating the reasons why great blues nest in colonies. Some theorize that the birds and their nests are more secure from danger in such numbers. Others believe that since the herons don't mate for life, a large colonial gathering facilitates the annual task of selecting a mate. Still others feel that colonies permit great blues to share information about prime feeding sites. Perhaps a combination of all these advantages is responsible for the heron's colonial nesting habits.

A heron rookery happened to be the indirect cause of an incident that our family treasures—an episode in which the misidentification of the great blue heron took a new, unexpected, and humorous twist. At the time there was a small rookery of great blues, visible from the highway but several hundred yards distant in a huge dead elm. Numerous spectators pulled their cars off the highway at this strategic spot in order to train binoculars on the nesting birds.

Our family was there one day, observing the herons, when a couple with two or three children joined the group of watchers. They happened to be standing next to us when one of the children inquired what the birds were. "Herrings," their father replied. "Blue herrings." All five of us heard this very distinctly, and we nearly choked with suppressed laughter. However, because we didn't want to diminish the father in the eyes of his family, we bit our tongues and refrained from any comment. Back in our car and safely out of hearing, however, we lapsed into paroxysms of mirth!

After mating, the female great blue lays three to six or seven eggs, with an average of about four. These hatch in four weeks, and the parents are thenceforward committed to feeding the voracious appetites of their rapidly growing brood. Considering that the heronry may be located miles from good food sources, the adults are often taxed to the limit to provide nourishment for the young herons. Back at the nest from a fishing expedition, an adult great blue feeds its nestlings by regurgitating partially digested food into their eager beaks.

About two months after hatching, the young herons begin to leave the nest and fly about. Soon they scatter and begin to hunt for food on their own. Now they must quickly learn the many critical lessons about feeding and foraging, or else perish—as many do. However, great blues that survive to full adulthood may live as much as twenty years in the wild.

In direct contrast to the great blue heron's tree colonies, whooping cranes are solitary ground nesters in marshy areas. There they construct mounds of bulrushes, cattails, and other marsh vegetation, with their nests in a hollow on top. Whereas the herons prefer trees, both for nesting and roosting, whooping cranes demand wide-open spaces where, one might assume, they can see predators approaching from a considerable distance. Indeed, trees are nothing but a hindrance to whoopers, because the construction of their feet is such that they can't even grasp a limb, let alone roost for any length of time. Where the great blue has an opposable back toe that enables it to grip a branch in combination with the front toes, the whooper's back toe is nonfunctional and almost vestigial.

As a sidelight, this anatomical feature of the whooper, which it shares with other species of cranes, may strike connoisseurs of Japanese art as a bit puzzling. Drawings and paintings from Japan frequently depict cranes roosting in trees. The answer to this little artistic conundrum is that Japanese cranes can't roost, either; the Japanese simply liked the effect gained by portraying them thus, and opted for artistic effect over scientific accuracy.

One of the most notable features of whooping crane behavior is their dramatic and highly ritualized mating dance. Male and female first face each other, then leap high off the ground with their legs and feet pointed at each other. Following this display, the pair bow toward each other, then repeat the entire performance over and over, often interspersing their movements with croaking calls.

After mating, the female whooper lays only two eggs. This number alone would represent a low reproductive rate, but that's only half of the crane's problem. Assuming that both eggs hatch, only the larger, stronger chick survives. Either the parents feed the larger one nearly to the exclusion of its smaller sibling, or the larger one pecks the smaller one to death or drives it out of the nest. Once outside the protection of the nest, of course, the chick quickly becomes prey for one predator or another.

This minuscule reproductive rate, combined with relentless hunting for its feathers and even more deadly habitat loss, caused the whooping crane population to dwindle steadily. Whoopers' numbers were probably never very high, but by 1941 only fifteen whooping cranes remained alive in the wild—all that stood between their species and extinction! Incidentally, some sources list the wild population of whooping cranes in 1941 as either fourteen or sixteen, but the official count at the Aransas National Wildlife Refuge in Texas, where the whoopers winter, was fifteen.

Efforts to save the whooping crane began at that point—efforts fraught with all manner of difficulties, not the least of which is the whooper's extraordinarily low reproductive rate. No one knew where the cranes nested, so until 1954 little could be done except give the remaining whoopers total protection and hope that they might eventually rebuild their shattered population, albeit at an excruciatingly slow pace.

Then serendipity struck. In that year, a pilot who was flying over Wood Buffalo National Park in Canada's Northwest Territories saw a pair of whooping cranes and what he believed might be a chick. He was right, and scientists now held the key to accelerated whooping crane reproduction.

After much study and debate, biologists decided in 1975 to remove one egg from some of the whoopers' nests. This had no adverse effect on whooping crane reproduction, since only one chick survives anyway. The pilfered eggs were then placed in the nests of the whooper's slightly smaller relative, the sandhill crane. This technique seemed to work well at first, as the sandhill cranes successfully hatched and raised the young whoopers. Then a major flaw became apparent.

Much of what whooping cranes do appears to be learned, rather than instinctive, behavior. Unfortunately, this happens to include mating behavior. Raised by sandhill cranes, the whoopers simply wouldn't mate and raise chicks when they reached maturity. Thus it was back to the drawing board for the scientists overseeing the whoopers' recovery.

Whooper eggs could readily be incubated artificially, but a major obstacle arose after the chicks hatched: How could they be raised without becoming imprinted on humans, which they would then regard as their parents? As biologists knew all too well from the sandhill crane experiment, cranes raised in this fashion would fail to reproduce.

The solution was to feed the whooper chicks using hand/arm puppets that resemble a crane's head and neck. Workers also donned crane costumes when caring for the chicks or carrying out other activities with them. The chicks and young adults are never allowed to see a human except under unpleasant circumstances, such as having a veterinarian catch and examine them. These techniques have been much more successful than the experiment with the foster-parent sandhill cranes.

As the number of whooping cranes in captivity rose steadily, in 1992 scientists began to release them into the wild on Florida's Kissimmee Prairie. Their goal was to create a nonmigratory flock containing 100 to 125 whoopers a year or more old.

Early releases were devastated by bobcats, which initially killed nearly two-thirds of the birds. In 1995, however, recovery experts changed tactics by switching from permanent to portable holding pens placed in safer habitat—low grass and freshwater marsh, where it's much more difficult for bobcats to stalk the cranes unseen until they're close enough to pounce. Once the cranes are properly acclimated, the pens are removed, leaving the birds on their own. With this change, bobcat predation on new crane releases has been reduced to a much more tolerable 30 percent.

The goal of one hundred whoopers age one and up in the Kissimmee flock now seems very feasible, for there are currently seventy-five of them, with the number steadily increasing as more are released. So far, none of those cranes has reproduced, but that all-important step seems to draw closer each year. Last year there were nests, but no eggs were laid. This year two pairs of whoopers nested, and each pair laid two eggs. Those eggs didn't hatch—possibly they weren't fertilized—but biologists are by no means discouraged. Whooping cranes, which mate for life and are long-lived, often take several years to settle into a routine of successful reproduction.

There are now 366 whooping cranes, 104 of them in captivity. In addition to the seventy-five wild whoopers in the Kissimmee flock, there are now 183 in the migratory Aransas flock—the original flock that was once reduced to only fifteen birds. This flock is growing at the rate of about 4 percent a year. There are also four remaining whoopers from the unsuccessful experiment using sandhill crane parents.

Biologists would also like to start another migratory flock—an extremely difficult task, since migration is a learned behavior in whooping cranes. Experiments are under way with both whooping and sandhill cranes to use ultralight aircraft as a means of leading young cranes to migrate.

With their present large population well distributed throughout North America, great blue herons appear to be thriving in most areas. Still, biologists note that there are at least three things that could reduce heron numbers in the future.

The first, ironically, is the return of the bald eagle, itself only recently taken off the endangered species list. Although the eagles only occasionally kill adult great blues, they prey on their chicks. More serious is the fact that they often frighten parents off the nests before their eggs hatch; as soon as a heron's nest is left unguarded, hordes of crows and other predators swoop in to seize the eggs. Although eagles thus take a direct and indirect toll of herons, it seems doubtful that the big raptors, which require a large territory, will become so numerous that they'll do serious harm to the herons. After all, they previously coexisted successfully for countless millennia.

The second concern stems from the great blue's ability to catch fish. As aquaculture has expanded dramatically in the past few years, great blues have discovered that these fish-filled ponds are pure largesse for a hungry heron. Naturally, owners of these commercial fish-raising ventures aren't thrilled at the sight of their profits vanishing down those elongated gullets, and so they've sought and received depredation permits to shoot as many as four thousand of the offending herons annually (great blue herons are otherwise strictly illegal to kill). While this number is by no means sufficient to harm the heron population, scientists are keeping a wary eye on the upward trend in depredation permits in case it becomes a problem.

The third concern—and by far the most worrisome in the long run—is increasing human intrusion near and in heron rookeries. Suburban sprawl, timber cutting, wetland drainage, development, and other human disturbances are slowly but surely nibbling away at this critical nesting habitat, which rarely has any legal protection.

Still, there is one encouraging sign in this regard: On Bloodsworth Island, an active U.S. Navy bombing range in Chesapeake Bay, great blues are happily nesting on wooden towers studded with nesting platforms. If great blues elsewhere display this sort of adaptability, it bodes well for the species.

Herons have been around for a long time—at least 14 million years, according to the fossil record—and the great blue heron, or a remarkably similar counterpart, existed nearly 2 million years ago. Despite the concerns already noted, the great blue seems destined to remain abundant.

Although the whooping crane's prospects grow brighter year by year, this magnificent bird is by no means out of danger. As a race, cranes are very ancient, and the earliest crane fossils date back roughly 50 million years to the Eocene epoch. By way of comparison, that's only about 15 million years after the extinction of the dinosaurs. Clearly, cranes have evolved in a way that's made them survivors over an immense span of time, but humans have recently added a level of stress never before experienced by these great birds.

Now, after nearly sixty years of painstaking work by dedicated biologists, the restoration of the whooping crane to its rightful place seems likely. Still, until at least another two flocks of successfully reproducing whoopers are established, it's too early to declare victory in the struggle to save this splendid crane.

Northern raven *(top);* American (common) crow

13

An Entertaining Pair: The Crow and the Raven

MYTHS

- A crow can imitate a human voice better if its tongue is split.
- Ravens are a great threat to calves, lambs, and other domestic livestock.
- Crows can count up to a certain number.

THE AMERICAN CROW *(CORVUS BRACHYRHYNCHOS)* AND ITS LARGER RELATIVE, THE RAVEN *(CORVUS CORAX)*, ARE LARGELY IGNORED AND FREQUENTLY DESPISED. That's very much our loss, for few birds will reward study and observation as much as these intriguing, astonishing, and sometimes very comical birds.

Of the two, the American crow is by far the more common and widespread. Three other species of crows, incidentally, are also found in the United States, though in much smaller numbers: the fish crow *(Corvus ossifragus)* and the Northwestern crow *(Corvus caurinus)* aren't nearly as widely distributed as the American crow, and the Mexican crow *(Corvus imparatus)* barely enters the United States in southern Texas.

The American crow inhabits most of the lower forty-eight states, as well as Canada north to the latitude of the junction of James Bay and Hudson Bay. Its larger relative, the common raven, is more northerly. It inhabits virtually all of Canada and Alaska, including their far Arctic reaches; New England; parts of a few other northern states; the Appalachian Mountains; and a broad swath from the Rocky Mountains to the West Coast and south into Mexico. It also appears that the raven is steadily increasing its range.

Telling crows and ravens apart where their ranges overlap can be difficult and may seem like a daunting task to the uninitiated. Size isn't a very reliable criterion, despite occasional wildly exaggerated accounts of the raven's size. A recent news item, for instance, called ravens "huge birds" and claimed that they're three times the size of crows. If that were true, the raven would indeed be huge—nearly four and a half feet long, with a wingspan of well over six feet!

Although the raven is somewhat larger than the crow, the differential isn't great: a big crow and a small raven are nearly identical in size. Despite the fact that size alone can't always distinguish between crows and ravens, especially at a distance, observers needn't despair: there are at least four or five other characteristics that singly, or especially in combination, will usually differentiate the two quite easily.

The tail, if it can be seen clearly at close range or silhouetted overhead, is one of the best identifying features. The crow's tail is square across its outer end, whereas the raven's is distinctly wedge-shaped at the rear. The beak is also useful for identification at short range, for the raven's beak is much heavier and more powerful than the crow's.

A third means of distinguishing crows from ravens is their calls. Although the two are closely related, their voices are much different: both have a very large repertoire of calls, some of them somewhat similar, but their most common and widely heard sounds are quite different and very useful in distinguishing between the two.

The crow's basic call is the familiar caw, often repeated over and over, but with different inflections and varying degrees of intensity, depending on the type of message that the crow wishes to convey. Alarm calls when an approaching human is sighted, for example, or the calls when crows are mobbing a hawk or owl, are far more intense and frantic than the caws that routinely go back and forth between crows. Crows also use numerous other calls, such as rattles, clicks, and various other sounds. However, these are heard far less often than the ubiquitous caw, at least partially because they don't carry as far.

Ravens, on the other hand, commonly use a variety of distinctive sounds. A guttural croak, a loud *rrrawwkk,* and a *quork* are among the most common. However, the most unusual raven call is a loud, ringing *goink* that sounds to me much like someone pounding on a huge wooden xylophone or gong, although others think it sounds more metallic. Ravens also gurgle, make noises which, at a distance, sound like people talking or laughing, and produce a variety of other sounds. It also appears that there are regional variations in raven calls, sometimes described as dialects.

One time-honored myth holds that a crow can be taught to talk if its tongue is split. This is nonsense. Crows are mimics, and have some capacity for imitating the human voice, but it falls far short of that displayed by birds such as parrots and mynahs, and a split tongue is of no help whatsoever.

Ravens do better in this respect, although that seems to depend a great deal on the individual raven and its inclination to imitate the human voice. For example, one scientist thought that it would be fun to teach a raven to say "Nevermore," as did the raven in Edgar Allan Poe's famous poem. The scientist succeeded—but it took him six years in the process!

On the other hand, there are credible reports of ravens performing amazing feats of mimicry. Scientist Bernd Heinrich has done outstanding work with ravens over the past fifteen years and contributed enormously to our understanding of these fascinating birds. In his book *Mind of the Raven,* Heinrich cites reliable sources for such things as a raven imitating radio static, a motorcycle being revved up, the sound of flushing urinals, and—most astonishing—an imitation of a demolition expert saying "three, two, one," followed by a reasonable facsimile of a dynamite explosion.

Although ravens clearly have the ability to mimic many things, including the human voice, their performance seems to depend on the *desire* to do so. Therefore, anyone wanting a bird that he or she can teach to say various words would probably do better to acquire a mynah or parrot.

Flight is a fourth means of telling crows and ravens apart, for it's perhaps in flight that the greatest difference between the two species is displayed. Although crows are excellent fliers, their flight is mostly a steady, uninspired, and laborious-appearing flopping. That doesn't mean that they can't perform more intricate maneuvers, but these are normally limited to special circumstances, such as landing in a high wind or mobbing a hawk or owl. Given their ability to roll, dive, and glide, their rather prosaic flight under most circumstances may be more a matter of choice than of ability.

Ravens, on the other hand, are magnificent fliers that seem to revel in displaying their talent. They routinely soar and glide like hawks, but they far transcend even that exalted level of flight. Often they engage in aerobatics reminiscent of mock dogfights between exhibition airplanes. In the process they perform barrel rolls, tumble in a variety of intricate moves, and even drop objects and catch them again while in flight.

At other times they display enormous precision by locking talons in flight, with one raven bottom side up beneath the other. In what's perhaps their most spectacular maneuver, they fold their wings and plummet like a skydiver for substantial distances before suddenly "opening their chute" and whipping into yet another display of aerial prowess. These spectacular aerobatics never fail to evoke a sense of wonder and admiration for these magnificent fliers.

The raven's powerful wingbeats can yield still another clue to its identity. A crow flying overhead makes little sound with its wings, even at rather low altitude, while a raven at the same height is clearly audible. Countless times when I've been in the woods, a distinctive *whuffwhuffwhuffwhuff* has made me look up to see a raven passing overhead.

The corvids—crows, ravens, jays, and magpies—are generally thought to be quite intelligent, but scientific evidence has been hard to come by. Bernd Heinrich, however, has done numerous experiments with ravens and given them the equivalent of avian IQ tests. In his most imaginative experiment, Heinrich tied food to one end of a string and attached the other end to a raven's perch. To reach the food, the raven had to perform a complex series of tasks: pull the string up with its beak, use a foot to hold the loop of string firmly against the perch, and repeat the operation several times until the food was finally raised within reach of its beak.

Several of his captive ravens were able to solve this problem on the first try and subsequently repeat the procedure. Others took longer, but were still able to come up with the same solution. This is especially impressive because Heinrich had taken pains to ensure that these ravens had never so much as seen a piece of string prior to the experiment. In contrast to the ravens, Heinrich's two crows (an admittedly small sample) never did figure out how to get at the dangling food.

Does this test prove that ravens are intelligent? The notion of intelligence is, at best, a hotly debated one. Respected scientists have widely divergent views on the subject, and a grand academic brawl is almost certain to ensue if anyone has the temerity to declare that this action or that test is proof of intelligence—or even to define intelligence, for that matter. And if scientists can't even agree on a definition of intelligence, it's virtually impossible for them to agree on a means of determining how much of it a creature possesses.

Heinrich is a cautious scientist, understandably reluctant to place his head firmly on the scientific chopping block by declaring that ravens are intelligent. However, he clearly leans in that direction when he says that he believes they experience some level of consciousness that they use in making decisions: "Whether that is 'intelligence' is subjective; but according to most people it is." I certainly concur in that judgment.

Crows are also widely regarded as intelligent, and perhaps they are, again depending on one's definition. However, one of the supposed measures of crow intelligence—the ability to count the number of hunters in a group—is highly suspect. According to this oft-repeated tale, if crows see, for example, four hunters go into a patch of woods, and then see only three depart, they know that the woods are unsafe because a hunter remains there. Also according to the tale, above a given number the crows can no longer tell whether or not all the hunters exit from the woods.

There might be a grain of truth in this: crows could conceivably recognize instinctively that two hunters emerging from the woods don't look like as many as three or four. However, such incidents are anecdotal, subject to a great deal of exaggeration, and, even if true, contain so many variables as to make them totally speculative and unreliable. Beyond that, using such an incident as the basis for saying that crows can count is straining credibility to the breaking point.

Crows are extremely social birds, and during the winter months they often gather in huge roosts that may harbor close to a million members! In the daytime they travel as much as twenty miles from these roosts in order to find food. Crows in large numbers can be a serious threat to crops, and need to be controlled under some circumstances.

As many farmers and gardeners know to their sorrow, crows especially like sprouting corn seeds. This has put the crow into marked disfavor with many, and has led to the use of various devices—scarecrows and imitation great horned owls among them—in an attempt (usually unsuccessful) to keep crows away during the critical period while the seeds are sprouting. However, on an overall basis, crows are regarded as beneficial because of the vast quantities of destructive insects, such as grasshoppers and cutworms, that they consume.

Crows have no compunctions about eating whatever is available, and items such as insects and sprouting corn form only a portion of their diet. Carrion, particularly in the form of roadkill, is a major source of food for the crow, and it's a common sight to see these birds on a roadkill or waiting near it until traffic permits them to resume feeding. Fruit, garbage, birds' eggs and nestlings, baby mice and voles, and a variety of other small creatures are also important food sources for the crow.

Large numbers of crows have been killed in the past, and some still are, largely because of the damage, real or assumed, that they cause. This seems to have had little effect on the crows, which have remained abundant. Ravens are a different story, for they've been seriously persecuted at times, not only with guns but also with poison and traps.

This enmity toward ravens probably has at least two causes. The first has to do with feeding habits. Ravens aren't crop eaters, and hence are much more oriented toward meat than crows. They're noteworthy carrion eaters, and Heinrich postulates an extremely close relationship between ravens and large predators, such as wolves, polar bears, and cougars. It appears that ravens follow these predators when they're hunting, thus gaining a share of the kill. Unable to kill large prey, or even to penetrate its hide, the ravens depend on the big predators to open up their kill and provide access to the innards.

Ravens are also known to peck at the eyes of dying animals, including domestic ones, because, in the absence of large predators, that's all they can get at. Farmers and ranchers, observing this behavior around dying livestock, or ravens around the carcasses of recently deceased animals, have accused them of killing lambs, calves, and even adult animals. Although the ravens lack the capacity, in most instances, to kill large animals, they've often been blamed for deaths that are really attributable to other causes. Nonetheless, this belief has led to the persecution of ravens and payments to farmers for reputed raven damage.

In addition to eating carrion, ravens are omnivorous feeders. Garbage, large insects, frogs, birds' eggs, nestlings, mice, a variety of invertebrates, and other small creatures are prominent in their diet. Fruit, berries, and seeds are also included in season. With their powerful beaks, ravens are also substantially more predatory than crows on midsized prey, including animals at least up to the size of squirrels and cottontail rabbits.

A second reason why ravens have been viewed in such a dismal light is a bit more complex and difficult to explain. To put it in its simplest terms, however, it's because the raven, as a carrion eater, has been closely identified with death. Although it's impossible to prove, this may have something to do with the raven as a scavenger of human bodies on countless battlefields down through the ages.

Certainly the raven hasn't always been viewed in such a negative way. Inuit and Native American legends honor the raven, and the Bible gives it excellent press. Genesis, for example, says that Noah sent forth the raven from the Ark, and the bird went to and fro until the waters were dried up. Likewise, in I Kings, ravens brought the prophet Elijah "bread and flesh" to sustain him.

Somewhere along the line, however, this favorable view of the raven broke down rather badly. Edgar Allan Poe's portrait of the raven as the herald of doom and despair is undoubtedly the best-known literary example of the raven's unsavory reputation, at least in the English-speaking world. It's by no means the only one, however. In the early 1700s, John Gay penned the lines, "That raven on yon left-hand oak (Curse on his ill-betiding croak!) Bodes me no good." And Sarah Helen Power Whitman wrote of the raven in this fashion: "Raven from the dim dominions / On the Night's Plutonian shore, / Oft I hear thy dusky pinions / Wave and flutter round my door. . . ."

Fortunately, a more enlightened view of the raven now seems to prevail, at least in most quarters. With protection, both raven numbers and the range of the bird have increased, and it seems likely that this trend will continue.

Added to the raven's other interesting feeding habits is their caching behavior. Ravens commonly eat part of their food and cache, or hide, some of it in various locations. The usual behavior on bare ground is to dig a little hollow, put a piece of meat in it, and then cover it with leaves or grass. In snow, the raven simply buries the choice morsel and lets the snow fall back to cover it. Heinrich's studies have revealed two particularly fascinating aspects to this caching behavior.

First, ravens have excellent short-term memory both for the precise location of their own caches and for the places where they've observed other ravens caching food; however, their memory for caches is poor after two weeks and almost nonexistent after a month. Second, ravens will make false caches—that is, apparently create a cache, but deposit no food in it. This behavior seems designed to reduce pilferage of caches, especially by subordinate ravens, which frequently attempt to steal the food cached by more dominant ones.

Ravens are far less gregarious than crows. Often solitary, they also commonly travel in pairs or very small groups, and are rarely found in large assemblages except in special circumstances, as when numbers of them gather around a carcass. That's not to say that ravens are antisocial birds, though. Indeed, Heinrich has documented behavior indicating that ravens may have rich social interactions among siblings or members of small groups, and I personally observed a group of them interacting in the following incident.

One day in early summer I heard what were distinctly raven calls emanating from the lower part of one of our pastures. I thought little of it at first, but my curiosity grew as the uproar became more and more prolonged. Even more curious was the fact that the ravens seemed to be stationary. Finally, after additional time had elapsed, I walked down to the pasture to investigate.

There, close to the woods, I found five ravens relentlessly harassing a thoroughly bewildered young red fox. No matter which way it turned, the ravens flew at it to drive it in another direction, herding it back and forth with great effectiveness. I watched for a few moments, then moved a bit closer, at which point the ravens suddenly broke off their chase and flew away. The young fox, as yet unharmed, promptly scampered into the woods.

Just what did this mobbing behavior signify? It seemed clear to me that the ravens were cooperating in this venture, whatever its purpose. Did they intend to kill the fox, or were they merely enjoying a bit of bullying? From their extremely aggressive behavior, I speculate that, given a little more uninterrupted time, they would have completely exhausted the fox and ultimately killed it.

There are other unanswered questions, as well. Were the five ravens a family, or were they merely a group that had assembled temporarily to attack the fox? If not family members, were they ravens that recognized one another and associated or cooperated with each other from time to time? Clearly there is still much to be learned about the behavior of these complex birds.

One other particular facet of raven and crow behavior that I've been able to observe extensively is their inordinate fascination with golf balls. We hit golf balls off our patio into the field below the house, and that has generated a number of incidents involving both crows and ravens. Likewise, we've experienced some unusual encounters between crows and golf balls on the golf course.

On several occasions we've watched ravens below the house, picking up golf balls and carrying them about. Sometimes they've pecked at them briefly, and other times not at all. Twice we've watched a raven walk about from ball to ball, picking them up and dropping them. That performance was immediately followed by quite different behavior: in each instance, the raven gathered dried grass with its beak and covered up a number of the balls. We were able to observe this activity in great detail through binoculars, since the ravens were only a hundred yards from the house.

The astonishing thing is that although I marked the location of these covered golf balls with great care, using a variety of landmarks, I had the devil's own time finding the balls—and some I never did locate! In each instance, the raven took only a moment to cover each ball, yet the job was done so cleverly and thoroughly that the ball became virtually invisible.

At another time, a raven "played" with several of the golf balls—then picked one up in its beak and flew straight to the broken top of a large dead maple along the side of the field. We were watching through binoculars and clearly saw the raven deposit the ball in the hollow in the broken top before taking wing again.

What these ravens were exhibiting was, of course, caching behavior. However, *why* they wanted to cache golf balls is another question. The obvious inference is that the ravens thought the round, white balls were some type of egg. Still, they had pecked at the balls, carried them about, and otherwise manipulated them with their beaks. It's difficult to believe that these seemingly intelligent birds would be fooled for very long into thinking that a golf ball was an egg.

It seems more likely that the egglike shape and appearance of golf balls trigger an innate response in the birds. Even though the ravens "know" from experience that the golf balls aren't eggs, the balls nonetheless arouse their curiosity and elicit a degree of possessiveness that leads them to cache these strange objects.

In view of this demonstrated fondness for golf balls, I wasn't surprised to learn that a large number of ravens disrupted a championship golf match in Iceland by swooping down and pirating balls. In fact, they did such a thorough job that the match had to be moved to another location! Neither does it seem strange that at the Yellowknife golf course in far northern Canada, close to the Arctic Circle, ravens make off with between two hundred and three hundred *dozen* golf balls a year just from the driving range.

Crows also have a thing about golf balls, though it takes a slightly different form from that of ravens. On several occasions we've seen crows pick up a golf ball or two, but they seem to lose interest in them faster than the ravens do. Moreover, crows don't exhibit the raven's caching behavior, so they made no attempt to hide the balls.

On the other hand, crows have long had a reputation as mischief-makers because of their propensity for collecting a wide variety of baubles, especially bright or shiny ones. On one particular occasion my curiosity was piqued by a black object partially hidden in the foliage of a tree along the edge of the field. Wondering whether it was a crow, a raven, or a black cat, I put the binoculars on it.

At that moment the crow—for that's what it proved to be—flew down into the grass and probed with its beak for a few seconds. Then it took flight, holding a shiny white golf ball securely in its beak. The last we saw of the crow, it was disappearing into the distance, the ball still firmly clamped in its beak. It probably was transporting the intriguing object back to its nest. Along the same lines, I recently found a golf ball several hundred yards deep in our woods, where it seems likely that another curious crow dropped it.

At least in this area, crows seem to delight in living around golf courses. No doubt the admixture of woods, brush, and mowed areas provides good opportunities for nesting and feeding, plus there are certainly tidbits of human food left here and there as an added attraction. Because of this, we frequently see crows at close quarters while playing golf, largely because these golf-course crows become quite tolerant of humans and let us approach them far more closely than is typical.

A number of amusing incidents involving crows and golf have been the result. On one occasion our older son and I were playing a hole where the fairway adjoins a steep, grassy bank. At the foot of the slope were two crows, facing each other across a golf ball. First one would pick up the ball and drop it, and then the other would do the same. This action went on for quite some time, until we eventually tired of watching it and proceeded with our game. The crows, on the other hand, seemed to be enjoying this activity and were still at it when we departed. How long they continued this unusual behavior is anyone's guess.

On another day I happened to be playing alone, quite early in the morning. By some miracle, I managed to hit a long, straight drive up the middle of the fairway on the first hole. I hadn't noticed a crow feeding in that vicinity when I teed off, but the ball landed near it, and the startled crow, with a loud exclamation, flew up into a nearby tree. I paid no more attention and trudged up the hill toward my ball. When I arrived at the spot, however, there was no ball to be found, no matter how hard I looked. Neither was the crow in the tree where I had seen it land. Although I have no proof, I'm certain that the crow flew back down and made off with my ball.

Most bizarre of all, though, was a golf-course incident that involved crows, or a single crow, but no golf balls. Our two sons, a friend, and I made up a foursome. All four balls landed on the green, and just as we walked onto it to begin putting, *plop*, a small pebble less than a half-inch in diameter landed on the green near us. We looked up to see a crow, only a few feet above our heads, just departing the scene. We remarked on the incident and commenced putting.

A moment later, *plop*, and the same scene was repeated. A minute or two later, *plop*, and a third pebble descended. Finally, just as we had completed putting and were walking off the green, *plop*, a fourth pebble fell near us. It's a mystery whether it was the same crow that dropped all four pebbles or whether different members of the same family were involved, although I suspect it was the former.

A far greater mystery is why the crow or crows acted in that manner. It's difficult to conceive of any biological or evolutionary purpose that could possibly be served by this behavior. Anthropomorphism is exceedingly dangerous, but it's hard to avoid the notion that the crow, or crows, were doing it out of some sort of pleasurable impulse.

The reproductive lives and nesting habits of the two species differ considerably, which isn't surprising in view of their divergent social habits. As befits their gregarious nature, crows tend to raise their young in an extended-family atmosphere, not unlike that of a wolf pack on a very small scale. That is, younger, unmated family members help the older, mated pair to build a nest and feed the young.

Crows construct a large nest of sticks and twigs, lined with soft materials such as grass and feathers. Given a choice, they usually select a site in an evergreen tree, mostly high up but occasionally lower. The eggs, normally four to six, are greenish with dark brown spots.

The young crows leave the nest after about three weeks. Thereafter, for a considerable time, they beseech their extended family to feed them, importuning them with reedy, pathetic-sounding little caws that resemble what one might imagine are the last, feeble utterances of a dying crow. Although their parents and other family members respond to these pleading calls to some degree, they gradually force the youngsters to fend for themselves, and their calls grow more adult in nature.

Ravens may mate when they're as young as three years old. On the other hand, some may not mate until they're as much as seven years of age. This behavior makes more sense when longevity is taken into account; ravens can live for as much as fifty years, although few in the wild survive to such an advanced age.

Ravens, unlike crows, have generally been considered to mate for life and to be completely monogamous. Like other animals that supposedly mate for life, however, the survivor will promptly mate again if its partner dies. Furthermore, subordinate males have been observed copulating with the female when the dominant male leaves the nest site. Evidently marital fidelity can't be taken for granted in either man or beast.

As might be anticipated from their more or less solitary nature, a raven pair builds its nest and raises its brood without the sort of help from an extended family that crows enjoy. Their preferred nesting site is a shelf beneath an overhang on the side of a cliff. Where suitable cliffs are absent, as they often are, ravens make do by placing their nest high up in a tall conifer.

The nest is made of sticks and is quite an impressive structure, roughly two and a half feet in diameter. The nesting hollow inside is about a foot wide and lined with a variety of soft materials, such as fur, shredded bark, and feathers. The eggs, numbering from four to seven, are a little smaller than a hen's egg. Their background color is greenish blue, but they're spotted and blotched to varying degrees with dark brown.

The nestling period is about forty days, which is unusually long. As soon as they leave the nest, the young ravens embark on a period of learning for the next month and a half to two months. During this span they follow their parents about, acquiring the knowledge of hunting and foraging necessary to survive. Then they begin to disperse, setting out to lead their own lives.

Birdwatching is reputed to be America's fastest-growing activity, and millions of North Americans, from the most dedicated birder with a life list to the casual observer of a backyard feeder, enjoy the pleasures of seeing a variety of birds. We particularly *ooh* and *aah* over brightly plumed birds, birds that sing sweetly, or raptors with fierce eyes, talons, and beaks.

Somewhere along the way, however, with the exception of dedicated birders, we tend to dismiss birds that are black and have raucous voices. Perhaps they seem dull and uninteresting, or perhaps they're so common that they fail to arouse our curiosity. The failure—and the loss—is on our part, not theirs. Crows and ravens are highly complex creatures in a number of ways, such as their social interactions and their vocal communications. They are also, in all likelihood, among the most intelligent of birds. The more we learn of these birds, the more complex and fascinating we find them.

14

The "Cat" that Isn't and Doesn't—and Its Relatives: The Amazing Weasel Family

MYTHS

- The "fisher cat" is a sort of cat that fishes.
- Fishers aren't harmed by porcupine quills.
- Fishers are large animals.
- Fishers kill porcupines by flipping them on their backs and attacking their unprotected bellies.
- Weasels are evil and kill for the sheer pleasure of killing.
- Weasels suck the blood of their prey.
- Martens feed mainly on squirrels.
- Our North American marten is the pine marten.
- Otters rarely go far from water.
- Otters kill many valuable fish.
- Skunks can't spray scent if picked up by the tail.
- Skunks are quick to spray if disturbed.
- Baby skunks can't spray scent.
- Female (or male) skunks have no scent.
- The skunk is a polecat.
- Badgers are rather placid and benign.
- The wolverine is virtually the devil incarnate.
- Oil spills have killed most sea otters in Alaska.

THE STOREKEEPER ASKED ME WHAT SORT OF ANIMAL HE HAD SEEN BOUNDING ACROSS THE ROAD IN FRONT OF HIM: HIS DESCRIPTION WAS A PERFECT MATCH

American marten *(top)*; fisher *(center)*; long-tailed weasel

FOR THE FISHER, BUT HE SHOOK HIS HEAD WHEN I TOLD HIM SO. "No," he said, "it was nothing like a cat." He was not an ignorant person, but he was the victim of a very common wildlife misunderstanding—that of the "fisher cat."

Far from being a cat, the fisher is a midsized member of the weasel family, among the most diverse and extraordinary groups of mammals.

Known to scientists as mustelids—the name ultimately stems from Latin *mus,* mouse, possibly because weasels are preeminent mousers—this family has exploited an extremely wide range of ecological niches. Their extraordinary diversity has required major adaptations in form and function, with the result that some members of the weasel tribe bear scant resemblance to anything we might think of as weasels.

Despite these differences, weasel family members share at least three major traits in addition to the fact that they are all carnivores. For one thing, they all have anal scent glands that can emit very unpleasant odors. The skunk has made a fine art out of this attribute, converting it into a potent defensive weapon, but other family members have equally unpleasant scents, though they lack the skunk's scent volume and efficient distribution system.

For another, most weasel family members utilize something known as *delayed implantation* in their reproductive cycles. This means that the fertilized egg doesn't attach to the wall of the uterus and begin to grow until months after breeding, and only a relatively short time before birth. Delayed implantation offers some major advantages, which will be explored more fully in the section dealing with weasels proper.

A third characteristic of the weasel family is a common gait that results in what might be called the two-step. In this method of locomotion, the tracks of front and hind feet are superimposed. The result is a track that looks as if it were made by two feet rather than four, one slightly to the rear of the other. Although some weasel family members use other gaits as well, this one is so typical that anyone who follows mustelid tracks very far is almost certain to see this distinctive pattern.

With these shared traits in mind, let's examine the unique qualities of the various weasel family members, for, despite the similarities, the differences between these creatures are striking. The weasel family is relatively old as mammals go; fossils of weasel ancestors at least 40 million years old have been found. In the immense interval since, evolution has worked its wonders. Form and function have become inseparable, so that the highly diverse shapes and sizes of the weasel clan are matched by the equally diverse ecological niches they occupy.

The Fisher

As already noted, the name "fisher cat" is a widely used and wholly inappropriate name for the fisher *(Martes pennanti)*. Although its glossy, dark brown outer fur looks almost black at a distance, and its tail is moderately bushy like that of some cats, it's hard to fathom why this large, forest-dwelling weasel has been associated with the feline race. Although proportionally thicker in the body than the super-slender weasel, the fisher is nonetheless elongated, short-legged, and possessed of a pointed, weaselish snout that looks anything but catlike. Likewise, the fisher's bounding gait hardly resembles that of a cat.

Even its correct name—fisher—is extremely misleading. Fishers certainly don't pursue live fish, nor would they be apt to eat fish at all except through the lucky—and infrequent—circumstance of finding a dead one. Whence the name, then?

In Old French, the polecat or fitchew—a European weasel relative—was called *fissel,* later *fissau.* This was gradually transmuted in English to *fitcher* and *ficher.* Ultimately, the term *fitch* was applied to the ferret, which is a domesticated strain of the polecat. The reasonable assumption is that European settlers in North America applied the name *ficher* to the unfamiliar animal that bore a cousin's resemblance to the familiar polecat. Incidentally, the Native American name of *pekan* or *pekane* is also used for the fisher on occasion.

As is the case with many other wild animals, the size of the fisher is often grossly overestimated. Far from being large, it is comparable in weight to a house cat, though longer. Like most others of the weasel tribe, fishers exhibit sexual dimorphism, which is the scientist's fancy term for a major difference between the sexes. In this instance, it refers to the fact that males are much larger than females. Thus a huge male can weigh up to twenty pounds, but that's unusual; adult males typically weigh ten or twelve pounds, females only five to seven.

The fisher is very much an animal of the northern forests. Most of its range is in Canada, though it is also found in the United States throughout most of northern New England, the northern Great Lakes region, the far northern Rocky Mountains, and parts of the Pacific Northwest.

Although the fisher eats a wide variety of food, its major claim to fame is as the only truly effective predator of the porcupine. How do fishers manage to kill porcupines with some degree of consistency when this feat mostly eludes much larger predators, including coyotes, wolves, cougars, and bobcats?

It's widely believed that fishers kill porcupines by approaching them from the front, then flipping them on their backs and attacking their quill-less bellies. But on close examination, this notion makes little sense, because it requires fishers to perform the extremely difficult feat of overturning unwilling victims as much as five times their own weight.

Research has shown that fishers use a different and much more effective method of attack. Extremely quick and agile, they dart in and out, repeatedly biting the porcupine around its unprotected face while dodging the potent slaps of its quilly tail. Eventually the porky succumbs to numerous bites or becomes so disabled that the fisher actually can roll it over with impunity and attack the unprotected belly.

The fisher's excellent tree-climbing ability offers a second major advantage, for it enables the fisher to attack porcupines when they're feeding or sunning themselves aloft. Despite these advantages, the majority of fisher attacks on porcupines fail because the porky protects its face in a den or a tree hollow, or between its paws.

The fisher's well-earned reputation as a porcupine killer has led some to the notion that fishers are immune to harm from quills. This is decidedly untrue: fishers are very good at avoiding quills, but occasionally one makes a mistake and gets a faceful. At best this is extremely painful, and at worst occasionally fatal.

Despite their fondness for fresh porcupine, fishers dine on a wide variety of food. Mice, voles, squirrels, chipmunks, and similar small creatures make up a major part of their diet, supplemented by grouse, hares, and the carrion of larger animals, such as deer. And despite being carnivores, fishers also consume a surprising amount of apples, berries, other fruit, and nuts.

How important is the fisher in controlling porcupines? For decades the state of Vermont, where fishers had long been absent, suffered from a plague of porcupines. Dogs routinely encountered them and subsequently paid a painful visit to the local veterinarian for quill removal. As noted in chapter 5, the big rodents also gnawed on everything salty from human sweat, including tool handles and canoe paddles—even outhouse seats! And damage to valuable timber trees, caused by porcupines chewing off their bark near winter dens, was rampant—and costly.

The state paid a bounty on dead porcupines for many years, but, as is virtually always the case with bounties, this measure proved ineffective. So serious had the situation become that in the late 1950s the Department of Forests and Parks began a program of putting poisoned apples (shades of Snow White!) in porcupines' winter dens.

Although the poisoning program was reasonably effective, Forests and Parks then began importing fishers live-trapped in Maine. These were released throughout the state, and within a few years the porcupine was back in a natural balance with its habitat.

Despite its success, this program was hardly noncontroversial. New in the experience of most Vermonters, the fisher engendered wildly exaggerated—even hysterical—fears. Some sportsmen blamed the fisher every time they failed to find grouse or snowshoe hares in their favorite covers, never mind that populations of these species are notably cyclical. Rumors of fishers weighing forty pounds and more were common. One person even phoned the Fish and Wildlife Department to report that a fisher had killed one of his heifers and then leaped over a fence with the heifer slung over its back—a feat beyond the prowess of even a full-grown cougar!

Letters to the editor warned of the dangers fishers posed to pets and small children. This was a gross exaggeration, to say the least. Although the fisher, like most predators, is an opportunist, and thus will certainly kill a house cat if it encounters one, it poses no threat to any but the very smallest dogs. As for attacking children, this is a ridiculous assertion; fishers simply do not attack humans, even very small children.

About twenty-five years ago, there was a great furor over the Vermont Fish and Wildlife Department's deer management policies. Some of the more vitriolic critics resorted to making anonymous (and completely false) allegations of moral turpitude and legal misdeeds against prominent department employees. As a result, I was eventually asked to chair a committee to investigate these charges.

In addition to deer management policies, the department was also being criticized for not trying to extirpate the recently reintroduced fisher. When our committee sought reasons for this, a prominent critic—who was decent enough not to hide behind anonymity—sent us a copy of a *National Geographic* publication that contained the striking assertion (as nearly as I can recall it), "Its green eyes glowing with hatred, the fisher attacks anything unfortunate enough to cross its path."

When we asked *National Geographic,* normally distinguished for its impeccable science, for the source of this astonishing description, they cited Ernest Thompson Seton. Alas for scientific accuracy, Seton, in many ways a remarkably accurate observer of wildlife, all too often lapsed into flights of anthropomorphism and melodrama. This instance is a prime example! All predators are fierce when it comes to killing their prey; they couldn't survive otherwise. The fisher is no more ferocious than most other predators, however, and indeed seems to be rather less fierce than many in its attitude toward humans.

Henry Laramie, a biologist with the New Hampshire Fish and Game Department for many years, asserts that fishers are very docile as far as humans are concerned. He once handled twenty-five fishers that were still in live traps; to prove this very point, he touched each fisher on the nose with his index finger as soon as he arrived at a trap. Not a single fisher bit him! On other occasions, Laramie had wild fishers loose in his car, running across his lap and shoulders; again, he suffered no bites. So much for the notion that fishers are incredibly ferocious!

The reproductive habits of the fisher, including delayed implantation, are quite similar to those of weasels (see below). The peak of their breeding season is in March, only about a week after the females give birth to a litter that averages three kits. The fertilized eggs remain free in the uterus for nearly eleven months; then they implant and begin to grow only about thirty-five days before the birth of the young.

The kits are born in a small cavity in a tree, usually about twenty feet above the ground. They are poorly developed at birth and don't even open their eyes for at least fifty days. From that point on, however, they mature rapidly and are able to be on their own by the onset of winter.

In summary, it can be said that the fisher has evolved as a medium-sized predator, primarily terrestrial but with substantial arboreal capabilities. Though frequently and unjustly maligned, it's an important part of the forest ecosystem within its range. This is especially true in regard to controlling porcupines. Further, the fisher is an extremely valuable furbearer. It deserves, and is generally receiving, management that assures its survival.

Weasels

Weasels have long had an unsavory reputation. We speak of a person as weaseling out of an agreement, or having a weasel face. There is also an underlying, though often unspoken, assumption that weasels are horrid, nasty little beasts, evil, cruel, and bloodthirsty.

In some ways this reputation is hard to account for. Viewed dispassionately, weasels, fierce little predators though they may be, present quite a different aspect: with their bright, beady eyes, alert and inquisitive faces, lithe movements, and relative fearlessness around humans, weasels can easily be seen as cute little creatures that are fun to watch.

More than likely, it was the weasel's occasional depredations on flocks of chickens and other domestic fowl, in a simpler time when most people lived on the land and raised their own chickens and ducks, that aroused much of this ire. And although one can understand the wrath of a farmer who found his flock of hens dead, the damage done to poultry by weasels was always minor compared to the good they did—and continue to do—in controlling mice, rats, and other rodents.

The notion that weasels are evil, cold-blooded creatures that kill for the sheer joy of killing stems from ignorance of their requirements for survival, fortified by a substantial dollop of anthropomorphism. Life is precarious for any wild creature, predator and prey alike, but especially for the weasel in winter. Consider some of the impediments to weasel survival during this harsh season.

First, the long, slender head and body, which enable a weasel to go most places where a mouse can go, are also very inefficient at preserving body heat; this problem is exacerbated by a thin fur coat and very little body fat. Second, the weasel is hyperactive, with a heartbeat in the hundreds of times a minute, so it takes a great deal of fuel to stoke its tiny furnace. Meanwhile, the pool of available prey steadily shrinks because young are rarely born during the winter months, while disease and predation take a constant toll.

Together, these ingredients form a recipe for weasel disaster—and in fact winter starvation is probably the major cause of weasel mortality. Against this formidable array of problems, evolution has programmed weasels with an effective survival strategy. This is known to biologists as "surplus killing," the trait that has given the weasel such a sinister reputation.

Because weasels live on the ragged edge of starvation, especially in winter, they kill as much prey as they can find; any surplus is cached and enables the weasel to endure at least a short period of unsuccessful hunting. Naturalists, incidentally, noted this caching behavior at least a hundred years ago.

Thus a weasel that slaughters a flock of hens in the coop is simply obeying an instinct. Its actions have nothing to do with taking pleasure in killing and everything to do with evolutionary programming that has enabled the species to survive.

Weasels customarily kill by a bite at the base of the skull, or close by in the neck. No doubt this fact, coupled with their penchant for surplus killing, led to the idea that weasels suck the blood of their victims, like minuscule vampires sprung to life out of some horror movie. A farmer finding his coop filled with dead chickens, uneaten but with tiny puncture marks in the neck, could be forgiven for assuming that the weasel was simply dining on blood.

Weasels have also been accused, from time to time, of sucking the contents out of eggs. In *As You Like It,* Shakespeare penned the words, "I can suck melancholy out of a song as a weasel sucks eggs." Unfortunately, the Bard was far off base with this analogy.

There are three species of weasels native to North America. In order of size, they are the long-tailed weasel *(Mustela frenata),* short-tailed weasel or ermine *(Mustela erminea),* and least weasel *(Mustela rixosa).*

The long-tailed weasel is found throughout most of the continental United States, except for Alaska, as well as a little of southern Canada. The short-tail's range is more northerly, encompassing even far northern Canada. It overlaps the long-tail's range in southern Canada, throughout New England, New York, and Pennsylvania, in the Great Lakes region, and in much of the area from the Rocky Mountains westward. The least weasel's range takes in Alaska and most of Canada except its far northern reaches, avoids New England and New York, then extends down the Appalachians and across the northern half of the United States to roughly the Rocky Mountains.

The long-tailed and short-tailed weasels are quite similar in appearance and habits, and they can be very difficult to tell apart except on close inspection, especially in their white winter phase. In their summer brown, however, short-tails can be distinguished by their white feet, which are lacking in the long-tails.

Long-tails are generally larger than short-tails (up to seventeen or eighteen inches long, including the tail, compared to thirteen inches for short-tails). Size, however, is not a very good criterion for quick identification; since male weasels are much larger than females, a large male short-tail can be as big as a small female long-tail. An average long-tail weighs eight ounces or less, an average short-tail less than half that.

Both species have a black tip on the tail. Except for this tip, the short-tail turns white in winter throughout most of its range. This, of course, is the famous ermine that long graced the robes of royalty. The long-tail also turns white in the more northerly parts of its range, where there is usually snow during the winter, and the white pelts with black tail tips of both species are considered ermine in the fur trade.

The white winter coat makes eminent good sense in an evolutionary scheme. Weasels are the target of numerous predators—hawks, owls, foxes, coyotes, and domestic cats among them—and the weasel's white fur against a snowy background provides excellent camouflage.

But what of the black tip on the tail, which would seem to defeat the weasel's camouflage? Scientists long puzzled over this seeming anomaly.

Then a researcher named Roger Powell wondered if perhaps the black tip of the tail drew the attention of predators away from the weasel's body. Using captive hawks trained to attack fake weasels "running" across a white background, he found that the hawks repeatedly attacked the black tip of the tail, rather than the "weasel's" body. When the black tip was eliminated, the frequency of successful body attacks increased.

Although long-tailed and short-tailed weasels are very small, they are veritable giants compared to the aptly named least weasel. The smallest of the world's true carnivores, this minuscule predator is only six and a half to eight inches long, including tail, and may weigh barely more than one ounce! It turns white in winter throughout all but the most southerly portion of its range. Curiously, however, its tail lacks a black tip; biologists believe that with a tail only one and a half inches long or less, a black tip would draw attacks too close to this tiny weasel's body.

Starvation and predators are by no means the only dangers confronting weasels. Although they're predominantly predators of mice, voles, and other small prey, weasels also attack larger animals such as rats and rabbits. Even the most peaceful of animals will defend itself with desperate courage and energy when cornered; hence weasels sometimes suffer fatal wounds when they attack prey many times their own size and weight. Incidentally, although weasels have excellent eyesight and hearing, they often use their keen sense of smell to track their prey.

Delayed implantation, already mentioned prominently, has some major advantages for a predator with the weasel's habits. In order to understand these advantages, let's start with the birth of a litter of kits, normally in April. There are usually six or seven kits, but litter size can vary widely.

The kits are blind, nearly hairless, and extremely tiny—premature by the standards of most mammals. This arrangement is vital to the mother, who must retain her slender outline and light weight in order to hunt successfully and survive throughout her entire pregnancy.

Breeding occurs three months later, in July or early August. The male comes to the den and woos the female with presents of dead mice and other prey; this allows the female to spend time breeding, rather than hunting.

Now comes the most unusual part—one that clearly demonstrates what a great advantage delayed implantation is to weasels. Not only does the male breed the mother, but her daughters as well! Because the fertilized eggs simply float around for about eight months before attaching to the uterus walls and starting to grow, the little juvenile females have time to grow to adult size before the fetuses develop within them. Since the average life span of weasels is so brief—about eighteen months—this breeding strategy ensures ample numbers of young to perpetuate the species.

By virtue of one of evolution's quirks, the least weasel is one of only two North American mustelids that don't utilize delayed implantation. Perhaps this is because the tiny predator breeds at various times of the year and may have more than one litter annually. Instead, the least weasel has a gestation period of thirty-five days; however, the kits develop with enormous rapidity during the final few days before birth, thus sparing the female from carrying a heavy load while hunting during most of her pregnancy.

Weasels seem to have little innate fear of humans. In fact, in their pursuit of mice they'll occasionally enter people's houses and remain there for some time, and they commonly search and inhabit barns, sheds, and other outbuildings.

I've been fortunate enough to observe this behavior on a number of occasions. In the most recent episode, I had skinned a deer in our shed a day or two before, and draped the hide over a rack. When I entered the shed, a weasel slithered off, oozing through a crack with lithe, serpentine movements. Moments later it reappeared, inspected me for a moment with curious, beady eyes, then ducked out of sight again.

Gradually becoming bolder, it soon began to gnaw at pieces of meat still attached to the deer hide, and became almost oblivious to my presence. I went back to the house, grabbed my camera, and proceeded to take a number of pictures of the weasel from a distance of no more than five feet, while it largely ignored me. This was typical of the weasels I've observed in our outbuildings.

We generate our own electric power by photovoltaic cells, with a backup generator. One winter evening I stepped into the generator house to start the generator, turned on the light, and was startled when a weasel, in full ermine garb, suddenly appeared. Paying little heed to me, the handsome little creature darted here and there, evidently seeking the trail of a mouse or red squirrel.

From time to time the weasel zipped out through the vents at the back of the generator house—evidently to check the wood stacked in the adjoining woodshed—then reappeared so suddenly and unexpectedly that I could only reflect on the appropriateness of the old song "Pop Goes the Weasel." Again, I retrieved my camera and took a number of photographs at close range, while the weasel seemed undeterred either by me or by the camera's flash.

Incidentally, the original of "Pop Goes the Weasel," quite different from the version that I grew up with, was written by the nineteenth-century English poet W. R. Mandale:

> *Up and down the City Road,*
> *In and out the Eagle,*
> *That's the way the money goes—*
> *Pop goes the Weasel.*

As already noted, many predators feed on weasels. However, domestic cats seldom seem to eat weasels which they've killed. On at least three occasions I found weasels in their ermine coats lying dead on our barn floor. Obviously killed by one of our cats, they had no marks on them except a bite in the neck. My suspicion is that the weasels' strong scent, so characteristic of their family, caused the well-fed cats to turn up their fussy feline noses at such fare. Wild predators, of course, can't often afford to be so fastidious!

Although they are victims of a great deal of bad publicity, much can be said in behalf of weasels. They're an important means of controlling mice, rats, and other small rodents, and this more than compensates for their occasional raids on poultry. Indeed, although weasels certainly do kill poultry on occasion, much of the killing blamed on them has been perpetrated by rats, an occasional mink, and other predators.

The Marten

Although weasels can climb trees, they are primarily ground-dwelling predators. Fishers are agile tree climbers and readily ascend after porcupines and squirrels, but they, too, are primarily terrestrial. The smaller, lighter marten *(Martes americana)*, however, is capable of exploiting the treetops more fully than the fisher.

Martens are about two feet long, including the tail, and usually weigh either side of two pounds. This light weight enables them to pursue the little red squirrel through the treetops on branches far too small to bear the weight of the much larger fisher. Thus, although it spends the majority of its time on the ground—where, after all, the majority of its larder is found—the marten can be considered semiarboreal.

No doubt the sight of martens zipping through the treetops in pursuit of red squirrels has given rise to the idea that they subsist mainly on a diet of squirrels. The marten, however, like the fisher, eats what it can readily catch or find. Mice, voles, and chipmunks make up much of the marten's diet, but birds and birds' eggs, hares, grouse, large insects, fruit, nuts, and carrion are all fare for this versatile predator.

With its yellowish-brown fur, a muzzle longer and more pointed than the fisher's, and larger, more pointed ears, the marten's face looks almost fox-like. And like the weasel, the marten can have its engaging side, too.

My good friend Tim Jones saw this facet of marten personality in an encounter in Maine some years ago. He had shot a buck about a mile from his car. After field-dressing the deer, he carried his pack, rifle, and heavy jacket out to his car. When he returned and began to drag the deer toward the car, a pair of martens suddenly emerged from inside the carcass.

Sometimes coming within a foot of Tim, the pair alternately tugged at the deer and chirred and chittered anxiously in high-pitched voices not unlike that of a red squirrel. According to Tim, their whole demeanor clearly said, "Chitterchitter! My deer! My deer!" The pair persisted and followed for some distance before finally turning back to content themselves with the leavings.

When he returned a second time for the deer's heart and liver, the martens were even more perturbed. In high dudgeon, they squeaked, chirred, and chittered their alarm and displeasure at seeing their food supply further diminished: "Chirr, chirr, chitter, chitter! My gut pile! My gut pile!" Tim was utterly charmed and captivated by the whole episode.

Our North American marten is commonly called the pine marten, but this is a misnomer. The true pine marten *(Martes martes)* is native to Europe. As with other New World creatures, European settlers evidently named our marten after a similar European relative with which they were familiar.

Martens are very much creatures of coniferous forests, or of mixed forests with a strong coniferous component. It's not surprising, therefore, that their range is primarily in Canada and Alaska. In the United States, they're found only in northern New England, the northern tip of the Great Lakes states, the Rocky Mountains, and a strip down the Pacific coast.

Like fishers, martens usually den in tree cavities, though a hollow log will also serve. In typical weasel fashion, they reap the benefits of delayed implantation. Usually three or four kits are born in April or May; though tiny and premature, they're covered with fine, yellowish fur.

A close relative of the famed sable *(Martes zibellina)* of northern Europe and Asia, the marten is a much-prized furbearer. Eliminated from some portions of its former range by the land clearing and unregulated trapping of bygone years, this interesting weasel cousin is now being restored to its old haunts.

The Mink

Curiously, and uncharacteristically for most members of the weasel family, little in the way of folk myths and erroneous beliefs seem to have sprung up concerning the mink *(Mustela vison)*. Why this is so is anyone's guess. Although infrequently seen, even in rural areas, the mink is widespread and common. Its range encompasses all of North America save Mexico, the Southwestern United States, and the Arctic reaches of Canada.

Mink are almost the same size as marten, though they are very slightly shorter and heavier. The differences make sense: the marten's light weight enables it to leap along slender branches, while the mink's body, just a bit stockier, works well for one who spends much of its time in the water. Considered semiaquatic, mink function well in an aquatic environment, although they lack the adaptations of their big cousins, the otters (see below). Good swimmers, mink lack the necessary lung power for sustained dives, and usually are submerged for only five to twenty seconds. Moreover, the mink's eyes are incompletely adapted for underwater vision. Consequently, mink spend much of their time foraging along the edges of streams and marshes, both in and out of the shallows, rather than diving after fish.

Frogs, salamanders, crayfish, and other aquatic creatures make up a good share of a mink's diet, with an occasional fish thrown in for good measure. Muskrats are also a favorite prey of mink, which follow them into their houses or burrows and there dispatch their hapless victims. In addition, mink prey on baby ducklings whenever possible.

Although mink are at home in and under the water, they also spend much time hunting on land. Mice, voles, rabbits, birds, eggs, and similar fare are all dinner for the mink while on land. Poultry also suffer occasional depredation by mink, and more than one hen-coop massacre blamed on weasels can be laid at the door of its semiaquatic cousin! Like weasels, mink engage in surplus killing and cache food; they also emulate weasels in trailing by scent.

Despite their wide range, relative abundance, and considerable time spent traveling about on land, mink are seldom seen. This is because they are primarily, though not completely, nocturnal.

Although mink are usually found near water, they're by no means wedded to it. Last winter I happened to glance out one of our windows and was surprised to see a mink running along the edge of our woods. We live a good quarter-mile from water, and the mink was headed in a direction where no appreciable water can be found for at least another mile. This seemed to trouble the mink not at all as it unconcernedly bounded on its way through the woods.

On another December day in a snowy forest, I found the typical two-step tracks of what I at first assumed was a weasel. On closer inspection, however, the tracks seemed too large for a weasel. Although the tracks came from an area where there was no water for a long distance, I became suspicious. Sure enough, the tracks eventually led to a small brook, mostly frozen over after a spell of very cold weather. There the tracks ended at a hole in the ice, about the diameter of a golf ball, its perimeter slightly discolored and worn smooth by the track-maker's repeated use.

Although mink raised on fur farms come in various mutant designer shades, wild mink sport a rich, brown pelage, except for a white chin patch. Their dense coat of underfur is covered by the long, dark guard hairs which give them their sleek, glossy appearance. Despite the availability of ranch mink, wild mink remain highly prized in the fur trade; evidently nature can still do a better job than humans of equipping mink with a dense, glossy coat.

Like all of the mustelids, mink deposit scent from their anal glands. However, the mink's scent is particularly strong and unpleasant; many would rate it worse than the odor of skunk, although the mink can't spray its scent as a defensive measure.

Mink breed during a period of over two months, starting in February. Implantation of the fertilized eggs is delayed for about a month; then a pregnancy of roughly twenty to forty days ensues. The young—typically three or four—are born, blind and covered with fine hair, in a hollow log, an expropriated muskrat house, or a burrow or cave in the bank.

The River Otter

If many members of the weasel clan are viewed by some with loathing and disdain, the river otter *(Lutra canadensis)* manages to salvage the family honor. Whether in the wild or in a zoo or animal park, it seems that everyone loves otters! The reasons aren't difficult to ascertain. With its bewhiskered visage, comically turned-down mouth, and often playful nature, the otter seems downright appealing to humans.

Aside from these attractive qualities, the otter is worthy of attention because of the way in which it has mastered its mostly watery habitat. Flawlessly equipped for this role, otters are a superb example of evolution-

Mink *(top)*; river otter

ary engineering. The mink functions quite well in and under the water, but the otter simply revels in it.

Otters are the closest thing to seals that freshwater provides. Propelled by wide webbed feet, the otter is magnificently constructed for gliding easily through the water. Its fur is short and sleek, its body long and cylindrical; even its long tail, thick at the base but tapering to a pointed tip, seems almost like an extension of its streamlined body, rather than an appendage. Moreover, its eyes are well suited for underwater vision, and it has the lung capacity to remain submerged for several minutes at a stretch.

Anyone fortunate enough to observe otters in the water for any length of time can readily see how at ease they are there. With the utmost insouciance, they roll, dive, play, and generally disport themselves with little apparent effort. Thus otters can accurately be termed "heavily aquatic."

Otters feed primarily on a variety of aquatic prey, particularly fish. For that reason they are sometimes blamed for killing large numbers of trout and other valuable game fish. This is generally quite unjust. Except for special circumstances (a small, stocked trout pond, for instance), otters, like any predator, will usually kill the easiest prey that requires the least expenditure of energy. That often means so-called "rough fish"—suckers, dace, and the like—rather than swift, wary game fish.

Several years ago my older son and I were offered a rare treat—a graphic example of the otter's liking for fish. We were canoeing down a broad, slow stretch of a stream appropriately named Otter Creek. As we approached a huge elm tree that had fallen into the water, we suddenly saw movement ahead of us. My son was in the bow, and he whispered, "Otters!"

Sure enough, there were two otters, rolling and diving around the skeleton of the old elm tree. We quietly shipped our paddles and drifted nearer, when suddenly one of the otters, with a sizable fish in its mouth, clambered up onto the trunk of the elm. There it proceeded to devour the fish at leisure, and we were so close that we could even hear it crunch up the bones!

Anthropomorphism—attributing human thoughts and emotions to animals—is an easy trap to fall into; animals aren't human and don't think and react as we do, however much we sometimes like to think so. It's extremely difficult, however, to put any construction other than sheer playfulness on some otter activities, notably sliding.

When otters climb steep banks and toboggan down a muddy, slippery slide into the water, time after time, there appears to be no biological imperative behind it; it just seems to be a pleasurable activity. Likewise, otters traveling in snow will fold their front legs back along their sides, push off with their hind feet, and glide down the slightest slope wherever possible—and sometimes even push and slide on a level. Whether sliding down a bank or gliding on the snow, the hind legs trail behind the body after giving an initial push.

Because otters spend so much time in the water, and are mostly seen in or along the edges of it, most people think they're almost completely tied to water. When the spirit moves them, however, otters can be notable overland travelers, traversing several miles of dry land before reaching another body of water. In the process, they are adept at finding small mammals and birds, insects, and similar terrestrial food.

Years ago, when our two sons were quite small, they were playing in an abandoned sugarhouse on a neighbor's land. Suddenly a movement caught their eye, and they looked out to see an otter humping its way along through the forest, far from any water. More recently, on two separate occasions we've seen an otter crossing the field below our house. As previously noted, this is far from water of any consequence. In both instances, the otter slid down every little slope and hummock, leaving a trail that resembled that of a tiny toboggan about eight or ten inches wide.

Otters mate in March or April. Then, following ten or eleven months of delayed implantation, the young are born about a year after breeding. Although as many as five young may be born, litters usually consist of two or three. The den may be a hole in the bank, beneath a fallen tree, or in an abandoned beaver lodge.

Except for the Arctic regions, otters are found throughout nearly all of Canada, Alaska, and the continental United States, except for small portions of the Southwest. As polluted waterways have become cleaner, otters have returned to many of their old haunts and are a fairly common sight.

Skunks

If otters are a great favorite with humans, skunks most decidedly are not! "As popular as a skunk at a lawn party," is a common expression, owing to the skunk's ability to deploy a malodorous spray as a defensive weapon. Even the scientific name of the striped skunk, our commonest species, delivers this message with double emphasis: *mephitis* comes from the Latin word for a noxious stench, and the name *Mephitis mephitis* is likely to be fervently endorsed by anyone unfortunate enough to be on the receiving end of a skunk's ire!

Curiously, skunks have fared better in the world of cartoons than in the real-life opinion of many. In addition to the cute Flower in Walt Disney's movie *Bambi,* there is the excessively and irrepressibly amorous character Pepe le Pew.

The word *skunk,* like the common names of several other North American animals (e.g., moose, woodchuck) represents the white settlers' version of a Native American word—in this case the Algonquian *seganku.* During the early days of colonization, rules of spelling were considerably less precise than they are now; one early spelling of *skunk* by a Massachusetts colonist was *squnck.* According to today's rules of pronunciation, this spelling results in "skwunk," a variation so delightful that our family regularly uses it.

At first glance, the portly skunk, which normally moves at a most sedate pace, would seem to have little in common with such swift, slender relatives as the weasel, fisher, marten, and mink. Like all members of the family Mustelidae, however, skunks have twin anal scent glands. Unlike the others, though, which can only dribble a tiny bit of scent, skunks can evert their anal glands—about the size of a grape—and discharge a potent blast of an oily, clinging substance with a most disagreeable smell.

Folk myths about skunks are numerous, and almost all deal in one form or another with the skunk's ability to spray its scent. One of the commonest and most persistent is that skunks can't spray if their hind feet are lifted off the ground; supposedly, if one can only grasp a skunk's tail and hoist before the skunk unleashes its scent, the hoister is thereafter safe from the skunk's quite understandable displeasure. (A related myth holds that a skunk can't spray if you step on its tail.)

Canadian biologist Chris Heydon, who has worked extensively with skunks, gives heartfelt assurances that this myth is decidedly untrue! Indeed, striped skunks have been known to walk on their front feet and occasionally spray from that position—a behavior typical of their cousins the spotted skunks (see page 166).

The popular perception of skunks is that they're quick to unleash a blast of spray if disturbed. In fact, skunks are generally very reluctant to spray, and rarely react in haste unless danger seems sudden and imminent. Usually a skunk will face a perceived threat, arch its back, elevate its tail, stamp the ground with its front feet, and shuffle backward. Only then, if the source of danger moves closer, will the skunk use its chemical defense.

It's fairly uncommon for humans to be sprayed by a skunk. Indeed, skunks seem to be remarkably tolerant of humans and spray us only *in extremis.*

There are even credible reports of people who have stumbled over skunks at night without invoking mephitic retaliation.

One bright moonlit night, our older son decided to take a shortcut through the woods from the college library to his dormitory. At one point he leaped over a log and, to his horror, landed beside a skunk! He recalls thinking, "Oh no, it's all over!" but the skunk, with remarkable forbearance, did nothing. This is only one example of the tolerance that skunks frequently display toward human disturbance.

On another occasion my father-in-law, who kept a couple of cows, went to separate the cream from some milk. When he approached the old-fashioned, crank-operated separator, he was somewhat nonplussed to find a skunk comfortably ensconced beneath it. It became apparent after quite some time that the skunk had not the slightest intention of vacating the premises, so my father-in-law approached cautiously, separated the milk and cream, and departed, leaving an unperturbed skunk, still beneath the separator where he had found it.

There are limits to a skunk's tolerance for humans, however. Take the case of Old Floyd. Old Floyd was a hired man who worked on our farm long before I was born, but his legend lived on. Something had been stealing eggs en masse from the henhouse, and it was finally deduced that the culprit was a skunk.

Old Floyd was—unwisely—given the task of rectifying this problem, so he set a trap near the point where the skunk was evidently entering the hen house. Something of a commotion near the henhouse became audible the next night, so Old Floyd set out with a long, stout stick and a lantern (this was in the days before flashlights).

Now, Old Floyd was noted for his clumsiness, and soon after he left the house, a far bigger commotion ensued. When my grandfather went forth to investigate, he found that Old Floyd had managed to trip himself with the stick as he approached the skunk, for it was indeed the marauding skunk in the trap. There sat Old Floyd on the ground, stick and lantern beside him, and the skunk in his lap! Like Queen Victoria, the skunk was not amused and had demonstrated its displeasure to the fullest possible extent. Needless to say, Old Floyd was something of a pariah for a considerable period thereafter.

If skunks haven't been genetically programmed to regard humans as a major threat, the same most emphatically can't be said for cats and dogs—to the considerable discomfiture of countless pet owners. For example, our daughter's cat leaped out an open window one night and landed on or beside a skunk. Retribution was swift and impressive!

Depending somewhat on wind direction, skunks can spray about twelve feet. They can also spray two or three times without recharging their scent glands—and a "discharged" skunk is nothing to cozy up to, for it only requires about a half hour to recharge its scent glands.

Yet another myth about skunks is that either the female or the male, depending on whose version one hears, can't spray. While it's true that female skunks are slightly less apt to spray than males, they definitely can and will use their scent in self-defense. Besides, who can determine a skunk's gender until it's far too late?

One of the worst features of skunk spray is its uncanny longevity. Despite repeated baths in one or more of the recommended antidotes to skunk spray, such as tomato juice or vanilla extract, the odor keeps returning for weeks whenever it's damp or wet. This is because certain very persistent compounds in skunk spray break down in the presence of water to produce the characteristic skunk odor.

An especially interesting feature of skunk scent is that a small percentage of people actually enjoy it! Evidently this is because some individuals have different scent receptors and perceive many odors—not just skunk—very differently from most people.

Although one encounter with a skunk is enough to deter future attempts by most predators, there is one prominent exception. As previously noted, great horned owls, which, like many birds, apparently have little if any sense of smell, regularly swoop down and kill skunks. Indeed, these fierce raptors are probably the skunk's only important enemy, aside from humans.

"Polecat" is a common term for the skunk in the United States, but this is a misnomer. The true polecat (the name comes from the French for pullet-cat, which probably memorializes the animal's depredations on domestic fowl) is a large European weasel known for its strong, unpleasant scent. As in the case of the fisher, European settlers probably began calling the skunk by the name of the nearest European equivalent.

Skunks occasionally have very large litters (the largest, recorded in Pennsylvania, was eighteen), but litters usually range from four to eight. Except in the eyes of devoted skunk haters, few sights in nature are more appealing than a mother skunk with her brood trailing single-file like cars behind the locomotive on a miniature railroad. However, therein lies a certain element of danger, for some people believe the rumor that baby skunks can't spray. There is a grain of truth in this: baby skunks can't spray for about twenty-five days after birth. By the time they're out and about with their mother, however, they should be considered armed and dangerous!

Striped skunks are widespread and common throughout the lower forty-eight states, a little of northern Mexico, and much of Canada. With a maximum weight of about fourteen pounds, and an average closer to eight or ten, they're about the size of a house cat. Typically, the striped skunk is black, with a white stripe on the head that divides into two broad stripes along the back; these often rejoin to form a white stripe on the tail, as well. Although this is the commonest pattern, skunk markings often deviate from this considerably.

Skunks are primarily nocturnal, although they can often be seen at dusk and occasionally in the daytime. They are truly omnivorous and will eat almost anything remotely edible, as many homeowners can attest after finding a skunk in the garbage can, or their garbage strewn about indiscriminately.

Absent food inadvertently supplied by humans, skunks eat earthworms and grubs, nuts and berries, birds' eggs, carrion, small rodents, ears of corn low enough for them to reach, and many other things. We could always tell when skunks were roaming our pasture at night because the older, drier cow droppings (or, if you prefer, cow flops, pasture patties, or meadow muffins) had been overturned during the skunk's search for the abundant earthworms and insects beneath.

The striped skunk is active throughout the winter in the South, but is a semi-hibernator in the North. There it sometimes dens communally, presumably for warmth. Skunks den under buildings, beneath stumps, and often in the abandoned burrows of animals such as woodchucks.

Although skunks in the North will occasionally emerge from their dens on particularly warm winter days, they mostly remain inactive until about mid-February. Then the mating call stirs within, and skunk tracks in the snow and dead skunks along the roadside can be seen with increasing frequency, as skunks travel about looking for a member of the opposite sex. The mating period continues until about the end of March.

A fairly brief period of delayed implantation follows early breedings, though not the later ones, and the young are born in May. Blind at birth, baby skunks are usually about two months old (remember that they can spray after about twenty-five days) before they leave the den to accompany their mother on her nightly forays.

The spotted skunk *(Spilogale putorius)* is our other common skunk. Absent from the East Coast, the Northeast, and parts of the Great Lakes states, this skunk otherwise inhabits Mexico and most of the remaining lower forty-eight states. Far smaller than its striped relative, the spotted skunk ranges from less than a pound to slightly over two pounds. A handsome little animal, it's really more striped than spotted; the markings on its black coat consist mainly of a series of broken horizontal and vertical white stripes on its back and sides.

Most of what has been said about the striped skunk applies to the spotted skunk as well. However, there are three differences worth noting. First, the spotted skunk has the peculiar method of standing on its front legs whenever it discharges its scent. What evolutionary purpose is thus served remains a mystery, but the creature has survived the ages and prospered.

Second, the scent of the spotted skunk dissipates far more rapidly than that of the striped skunk. Whereas the chemicals in the latter's spray break down slowly in the presence of water to release the characteristic skunk scent, those of the former release their scent very rapidly.

Third, the spotted skunk will sometimes climb trees if danger threatens, although it rarely does so in the normal course of events. The striped skunk, on the other hand, is no tree climber.

Two other species of skunks barely reach into the United States, although they inhabit Mexico. The first is the hooded skunk *(Mephitis macroura),* which is found only in the extreme southwestern states. Intermediate in size between the spotted and striped skunk, it has a much longer tail than either. It may either be black, with a back that's mostly white, or all black except for two narrow white stripes on each side.

The hognose skunk *(Conepatus leuconotus)* is also called the rooter skunk. As the name implies, it roots for much of its food with an elongated, somewhat piglike snout that is naked for an inch or so on top. About the length of a striped skunk, it's less chunky and weighs only two to six pounds. Its entire back and tail are white, making it a very distinctive two-toned animal. Its range extends into about half of Texas, as well as much of Arizona and New Mexico and a small slice of Colorado.

Where other members of the weasel family have claimed a fairly specialized environmental niche, skunks of all stripes (pun intended) are generalists. With few enemies to bother them—at least more than once—they've evolved with a leisurely lifestyle well suited to their rather rotund bodies, slow gait, and notably unfussy appetites. Yes, skunks are generally regarded in an unfavorable light and can be a nuisance at times, but they're actually interesting and generally beneficial animals that devour a lot of potential pests. To anyone really familiar with them, it's difficult to escape the notion that they deserve better press.

The Badger

The badger *(Taxidea taxus)* is nature's steam shovel. Whereas other members of the weasel family have variously come to utilize the treetops, the land's surface, and the water's depths, the badger is fossorial, meaning burrowing or digging. Indeed, it is superbly equipped for subterranean life.

Badgers are found from Ohio, Indiana, Illinois, and the Great Lakes states westward throughout most of the remaining lower forty-eight. Their range extends southward into parts of Mexico and northward into the western Canadian provinces.

With its wide, flat body, very short legs, and grayish brown coat, a hunkered-down badger very much resembles a doormat, albeit one with a head sporting black and white facial markings. From two to two and a half feet long, badgers are sturdy creatures that weigh as much as twenty-five pounds.

In company with skunks, badgers seem unlikely relatives of the svelte, swift weasel, fisher, marten, or mink. However, they bear the unmistakable hallmarks of their kin, including the anal scent glands.

Badgers have developed several adaptations marvelously suited to such subterranean proclivities as tunneling after ground squirrels, prairie dogs, mice, and other small rodent prey; scooping out burrows for daytime sleeping; and digging themselves rapidly out of sight if danger threatens.

Foremost among these adaptations is a set of incredibly long, strong front claws. A full two inches long, they would do justice to an animal many times the badger's size. These are coupled with another adaptation, webbed front toes. Thus the badger, using its powerful front limbs, can rip out the soil with its great claws and hurl the loosened earth backward with its scooplike front feet. In earth suited for digging, a badger, clouds of dirt flying behind it, can dig its way out of sight in about two minutes!

A third adaptation is a transparent inner eyelid, called a nictitating membrane. This can be drawn across the eye when necessary to keep dirt out, enabling the badger to see even when digging furiously.

People who have never encountered a badger—if they think about badgers at all—tend to regard them as relatively placid and benign creatures. Those who have had dealings with badgers know better.

Probably much of this image can be traced to Kenneth Grahame's great classic, *The Wind in the Willows.* In the utterly charming world that Grahame constructed, Mr. Badger is a gruff but kindly individual, possessed of great wisdom and integrity. Badgers are gruff, all right, but kindly they are not! A nineteenth-century writer, John Clare, summarized the badger's disposition

Black-footed ferret *(top);* badger

with great accuracy: "When badgers fight, then everyone's a foe."

Unlike European badgers, which often live as an extended family in a warren of burrows called a *sett,* American badgers are solitary creatures except for a brief period of mating. In keeping with their hermitic nature, badgers are extremely displeased by any invasion of their rather outsized personal space, which they will defend aggressively.

If a human, a would-be predator, or even another badger approaches too closely, the response is usually a series of ferocious growls and hisses, backed up by an impressive show of teeth—and by all accounts, the sound levels produced by a badger angry at being caged are truly fearsome!

This display is no mere bluff. Badgers are noted for their courage, ferocity, and exceptional strength, as many larger creatures have learned to their sorrow. Badgers regularly whip dogs several times their size and not infrequently kill them.

Complicating matters for anything that attacks a badger is its thick, tough, exceptionally loose hide. If, for instance, a predator seizes a badger by the neck, the loose hide permits the badger to turn and savage its tormentor. As a result, adult badgers have almost no serious natural enemies, although wolves and cougars probably preyed on them at one time. Nowadays, automobiles represent the worst threat to badgers, which seem to be no more afraid of them than of other enemies.

Badgers, on the other hand, prey widely on many species. Besides a variety of rodents—a staple of their diet—badgers happily consume raccoons, armadillos, larvae of bees and wasps, snakes, lizards, birds' eggs and young birds, frogs, crayfish, and sometimes carrion. In addition, they will dig out the dens of foxes and coyotes and devour their pups. Thus does the image of the sedate, kindly badger break down!

Except for a mother with young, badgers seldom spend two days in a row in the same den. Instead, these mostly nocturnal mammals simply dig a new burrow wherever they find themselves at dawn. Although badgers are generally active throughout the winter, they have no qualms about fashioning a cozy burrow and staying there for several days to wait out a spell of especially bitter weather.

Badgers breed in late summer; as in most mustelids, implantation is delayed for several months, and the young aren't born until the following spring. The cubs, usually two or three per litter, are born blind and with very little fur.

By late spring, after being weaned and then introduced to meat brought by their mother, the cubs travel with her for several weeks, learning to forage. By summer, however, they either leave their mother voluntarily or are driven off to lead the mostly solitary lives that seem well suited to these markedly short-tempered animals.

The Black-footed Ferret

The black-footed ferret *(Mustela nigripes)* holds the dubious distinction of being the only endangered species among North American mustelids. Weighing approximately one and a half to two and a half pounds, this marten-sized weasel closely resembles the steppe polecat *(Mustela eversmanni)* of Eurasia. Mostly yellowish brown, it has a lighter face with a black mask across the eyes and forehead and a black-tipped tail. Further, as both the common and scientific names indicate (*nigri,* black, plus *pes,* foot), all four feet are black.

All other North American members of the weasel family seem to be at least holding their own throughout wide portions of their range, so how did this distinctively marked animal end up in such a precarious situation? The answer, in simplest terms, is overspecialization.

Principally nocturnal, the ferrets prey very heavily, sometimes almost exclusively, on prairie dogs; when these rodents disappear, so do the ferrets. Over a long period, several things have drastically reduced prairie dog numbers throughout the ferret's range.

First, much prairie was converted to agricultural uses unsuitable for prairie dog habitat. Second, many prairie dog colonies have been eliminated by poisoning, since the rodents were considered a nuisance. And third, prairie dogs began dying from sylvatic plague, which is the animal equivalent of bubonic plague—the infamous Black Death of the Middle Ages.

By 1985, only eighteen live black-footed ferrets were known to exist, and it was feared that they would die from distemper (actually, it was later found that sylvatic plague was a greater threat). At that time, a decision was made to live-trap them and try to raise a captive population; if their numbers increased greatly, it might then be possible to reintroduce some to the wild.

Fortunately the eighteen captive ferrets thrived and multiplied. As their numbers increased into the hundreds, efforts were made to reestablish them in the wild. Some attempts have failed, but the program has had good success in Montana and South Dakota. At present there are around 150 ferrets in the wild and about 350 in captivity. Although those numbers are encouraging, ferret experts caution that the future of wild populations is still very uncertain.

Besides its endangered status, the black-footed ferret is unusual in another way: together with the least weasel, it's the only North American mustelid that doesn't have delayed implantation. It does, however, have what is known as *stimulated ovulation.* This means that the female doesn't ovulate until breeding by the male causes her to do so. The two to five young are born in June and disperse as adults in August.

The U.S. Fish and Wildlife Service, under the Endangered Species Act, has done an outstanding job of bringing the black-footed ferret back several long steps from the brink of extinction. As with some other endangered species whose numbers are increasing, however, it's still too early to declare victory in the effort to save this important cog in the prairie ecosystem.

The Wolverine

If skunks have widely been viewed with disfavor, the reputation of the wolverine *(Gulo gulo)* has been far worse. When I was growing up, stories in outdoor magazines and tales in books about the far north portrayed the wolverine as possessed of a demonic hatred of humans, abetted by a fiendish cleverness—in short, a devil in animal form. More enlightened thinking, combined with scientific observation, paints a rather different picture.

The second-largest member of the weasel family, the wolverine can weigh as much as sixty pounds. Admittedly, it's a decidedly odd-looking creature; to paraphrase the famous description of the camel, the wolverine looks like a small bear designed by a committee.

With a bearlike head and body, the wolverine has a short, bushy tail that somehow looks like an afterthought. Its coat is thick and dark brown, except for a lighter band across the top of the head and a wide, yellowish stripe along each side; these stripes start at the front shoulders and merge at the tail, much like the stripes on a skunk.

The principal range of the wolverine is the far north, from Alaska across Canada and into that nation's farthest Arctic reaches. Wolverines are also found southward through British Columbia into Idaho and perhaps a few other pockets in the United States.

Unfortunately, not enough is known about remaining wolverine habitat in the lower forty-eight states. Money for wolverine research is scarce, and much needs to be learned about where they live and how many are left. What is known is that wolverines were once found much more widely in the United States, including the Midwestern and Eastern states. Ironically, there is

Wolverine

absolutely no evidence that wolverines ever inhabited Michigan, which is known as the Wolverine State.

The wolverine's diabolical reputation is based largely on a combination of tall tales and misinterpreted facts. The accounts of wolverine-as-devil usually go something like this:

A trapper runs his trapline from a cabin in the far northern wilderness. One day a wolverine discovers the trapper and instantly acquires a dire hatred of the man. The wolverine then begins checking the trapline ahead of the trapper, killing and eating the catch in order to drive the trapper away. Desperate, the trapper attempts to trap or shoot the wolverine, but to no avail; with fiendish cunning, the wolverine continually eludes him.

One day the trapper returns to his cabin to find that the wolverine, as further proof of its diabolical nature and all-consuming hatred, has broken into the cabin and eaten or destroyed the trapper's winter supplies. The trapper is forced to give up and return to civilization, the victim of a supernatural adversary.

No doubt there is considerable truth to at least some of these tales, but the trapper's interpretation of the facts are wildly inaccurate. The actual story is probably as follows:

The wolverine evolved as a denizen of an extraordinarily harsh climate. Lacking particularly good eyesight and hearing, this mammal is nonetheless wonderfully adapted to its home in the far north. These adaptations include a marvelously keen sense of smell, great endurance, and immense strength for its size; indeed, many consider the wolverine and badger to be, pound for pound, the strongest of all North American mammals.

The wolverine, to survive in such a harsh environment, became a hunter and scavenger in the summer, caching quantities of food against the long winter, and primarily a scavenger during the winter. Using its exceptional sense of smell, the wolverine could detect its caches and the remains of old wolf or bear kills beneath several feet of snow.

One day a trapper moved into territory inhabited by wolverines. He built a cabin and, like the wolverine caching food for the winter, stocked the cabin with enough staples to see him through until spring. Then, on snowshoes, he set his traps along a circuit covering several miles.

Along came a wolverine, which, curious about this strange scent, followed the trapper's trail. There it found animals immobilized in traps—a hungry wolverine's version of Easy Street. Since wolverines are far from stupid—though certainly not endowed with supernatural intelligence—the wolverine quickly learned that following the trapper's trail meant easy pickings, and frequently checked the trapline in the course of its peregrinations.

To the superstitious trapper, it seemed as though the wolverine was acting out of malice and trying to put him out of business. This notion was reinforced when the trapper, despite repeated efforts, failed to see the wolverine in order to shoot it, and the wolverine evaded traps set for it.

The final act came when the wolverine, particularly hungry that day, happened by the trapper's cabin. Its keen sense of smell detected food within, and, with its great strength, long claws, and formidable teeth, it was able to break through a door, shuttered window, or vent. Once inside, the wolverine tore open containers of flour, sugar, bacon, beans, and other goodies. After gorging itself and leaving the interior of the cabin in an unholy shambles, the wolverine went on its way.

When the trapper returned and found most of his winter staples wiped out, he had no choice but to leave—additional proof of the wolverine-as-devil-in-animal-form. Of course, in addition to genuine superstition, it was only natural for the trapper to embellish the story a little—well, maybe more than a little—in order to explain his way out of an embarrassing situation. Of such threads are the fabric of animal reputations often woven!

The reputed ferocity of the wolverine toward humans, and its supposedly supernatural cunning, is a good example of the mythology that has grown up around this creature. There are, for instance, several documented cases of people raising wolverines, which made rather docile pets. (Raising wild animals in captivity is only rarely a good idea, however, and should be left strictly to qualified experts!) Clearly the wolverine's ferocity toward humans, like the fisher's, has been greatly enlarged.

A cornered wolverine in the wild is quite a different proposition. In keeping with its bearlike appearance, the wolverine is reliably reported to have a most bearlike growl, quite out of keeping with its modest size. Combined with an impressive display of teeth, this is sufficient to make almost any animal or human think twice about coming closer. Nevertheless, packs of wolves are known at least occasionally to kill wolverines.

If much about the wolverine has been grossly exaggerated, its strength has not. For example, one wolverine in Alaska was observed dragging a Dall sheep carcass three to four times its weight for roughly two miles; numerous other feats of wolverine strength have also been noted by wildlife biologists. In particular, wolverines are notorious for chewing or breaking their way out of extremely strong, well-constructed cages, for its exceptionally powerful jaws, designed by nature for crunching up carcasses and bones while scavenging, can also wreak havoc with man's efforts to confine it.

Wolverines are notable travelers, constantly on the move as they scour their territory for food. Indeed, the speed and distance of their odysseys evoke awe even in sober scientists: a wolverine can run for miles without stopping, and may travel fifty to sixty miles in a single day, while covering a range of several hundred square miles.

Wolverines also have other adaptations to help them cope with their often brutal environment. Its outsized paws are a major asset while traveling in deep snow, and formidable claws enable wolverines to dig out prey, cached food, and the kills of other predators; these same claws also make wolverines good tree climbers.

The wolverine's coat displays yet another evolutionary gift. The dense, warm underfur is topped by long guard hairs that are tapered so that they shed frost and ice much more easily than the hairs of other animals. Natives of the far north long ago realized this and began using wolverine fur in parka hoods: when breath condenses into frost around the hood, it's easily brushed off the tapered wolverine hairs.

Except for a brief summer mating period, wolverines are solitary. Somehow, even in such large territories, males manage to find the females; then, after mating, they go their separate ways for the remainder of the year. Implantation is delayed until late winter, and two or three young, small and blind, are born in the spring. The natal den may be a deep crevice in the rocks or a space beneath a stump, log, or blown-down tree. The young wolverines are independent by fall, and disperse to seek their own territories.

While there is concern that the remnant wolverine populations in the lower forty-eight states may be disappearing, the wolverine seems to be thriving, at least for the present, in the vast reaches of the far north. It is a fascinating and unusual creature that deserves far more study and attention than it has heretofore received.

THE SEA OTTER

Largest of the weasel family in North America, the sea otter *(Enhydra lutris)* has gone the river otter one better and become, for all practical purposes, a fully aquatic marine mammal. Although sea otters occasionally come on land for brief periods, such as waiting out severe storms or to bask on the rocks, most of their lives are spent completely in and under the waters of the North Pacific.

Hunted relentlessly for their valuable fur, sea otters were brought to the edge of extinction, and at one time only about two thousand remained throughout their entire range. Now, after years of full protection, there are probably 150,000 to 200,000 of them—a remarkable comeback.

Like river otters, sea otters are much loved by everyone—well, almost everyone; Alaska crab fishermen and California abalone and sea urchin fishermen tend to loathe sea otters with a passion. According to sea otter expert David Garshelis, groups of otters, known as pods, can indeed severely deplete local crab populations, at least temporarily, in Alaskan waters.

Jack Ames of the California Fish and Game Department believes that sea otters were historically in balance with abalone, sea urchins, and other species on which the otters feed. Then, when the sea otters were virtually wiped out 150 or more years ago, the population of abalone, sea urchins, and other creatures prized by humans as food grew rapidly. Soon a thriving commercial fishery grew up around these species and came to be regarded as "traditional."

With the return of sea otters in substantial numbers during the second half of this century, the old balance between predator and prey, which didn't include commercial fishing by humans, was reestablished. As Ames says, a commercial abalone fishery simply can't exist in the presence of sea otters.

Sea otters commonly weigh up to eighty or eighty-five pounds, and a few specimens have reached one hundred pounds. Lacking the seal's insulating layer of blubber, sea otters survive the cold ocean water by virtue of the densest fur of any animal—100,000 hairs per square centimeter!

Nor is this extraordinarily warm, thick coat the sea otter's only adaptation for marine life. Very large lungs serve a dual function: they allow for extended time underwater, and they also permit the sea otter to float effortlessly on the surface. Indeed, sea otters actually sleep while floating on their backs.

In addition, sea otters have completely webbed, flipperlike feet that are perfect for swimming, though extremely awkward on land. Moreover, in another curious evolutionary twist, the hind feet are "backwards"; that is, the big toe on the hind foot is on the outside of the foot, rather than on the side toward the body. This peculiar adaptation may make it easier for the otter to swim on its back, a position in which it spends much of its time.

Besides sleeping on its back, the sea otter will often emerge from a dive with both a shellfish and a small, flat rock. Rolling onto its back, with the rock on its stomach, the sea otter next holds the shellfish in both front paws and pounds it against the rock to break the shell and expose the meat within. Certainly this is a fascinating use of a tool by this animal!

One popular myth, at least as far as Alaskan sea otters are concerned, is that oil spills such as the infamous *Exxon Valdez* disaster wiped out sea otters along much of the Alaska coast. It's true that sea otters are more likely to die from oil spills than are seals; the latter have blubber to insulate them, while sea otters, once their protective fur is matted by oil, succumb to the cold. However, as destructive as oil spills are, they don't reach into many bays and setbacks in the shoreline, where substantial numbers of sea otters avoid the oil and survive. Oil spills such as that from the *Exxon Valdez* certainly do great harm to the local population of sea otters, but don't eliminate them.

Male sea otters often congregate in large rafts or pods, some containing more than five hundred otters. Females also form pods, though substantially smaller ones.

Like most of their weasel family brethren, sea otters have delayed implantation. Unlike their relatives, however, baby sea otters are fairly large (about three pounds) at birth, with eyes wide open and a furred body. This makes perfect sense; blind, nearly naked young would have little chance of surviving in the frigid ocean.

In virtually all cases, the mother sea otter bears only a single pup. Although relatively few sea otter births have actually been observed because they happen so quickly, pups apparently may be born either on land or in the water. The pup sleeps cradled on its mother's stomach, while she floats easily and contentedly—a sight that has certainly helped endear the sea otter to countless people.

Despite their incursions against commercially valuable species such as crabs, abalone, and sea urchins, it's gratifying to know that sea otters are no longer in danger of extinction. They are unique and fascinating—an extreme example of evolutionary adaptation in a family that has specialized in it.

Just as the last piece of an intricate jigsaw puzzle, snapping into place with a faint but satisfying click, makes the image on the puzzle's face entire, the sea otter's mastery of the marine environment completes the picture of the weasel family's extraordinary diversity. Stemming from a common ancestor some 38 million years ago, family members evolved to utilize virtually every major type of habitat: trees, the land surface, the subterranean, freshwater, and the sea. Only the air—the exclusive domain of the bats among mammals—is denied to this amazing family!

15

Cute Isn't Tame: The Raccoon

MYTHS

 Raccoons make good pets.

Raccoons are extremely clean and wash their food before eating it.

MOST PEOPLE FIND RACCOONS *(PROCYON LOTOR)* EXTREMELY APPEALING, AND THE MOST FREQUENT TERM APPLIED TO THEM IS "CUTE." Cute they may appear, with their black masks and furry, ringed tails, but biologists and others who deal regularly with wildlife take a somewhat more jaundiced view of this animal.

Because of the cuteness factor, especially in baby raccoons, people often want to make pets out of them. Anthropomorphism strikes again: raccoons are cute, so they must be nice, cuddly little animals that would make good pets. This is a big mistake! The adorable little raccoon that someone picks up and takes home frequently grows into a vicious and highly destructive nuisance.

Despite their visual appeal, raccoons are fierce predators and tend to have nastier dispositions around humans than do many other predators. Yes, some people have successfully made pets of raccoons (almost always illegally), but far more often the "pet" baby raccoon turns into a nasty-tempered creature that bites, scratches, and raises havoc with household possessions.

Much more serious, however, is the threat of rabies; in fact, one of the major rabies strains is called the raccoon strain because raccoons are its primary carrier. Rabies frequently has a long incubation period of several months' duration, so the cute little raccoon that someone tries to raise as a pet may become rabid weeks or months later.

Raccoon

A Massachusetts woman serves as Exhibit Number One in this regard. Unwilling to see three little raccoons euthanized, she ignored all warnings and raised them illegally. One died three months later and tested positive for rabies. As a result, nineteen people had to receive rabies shots at a cost of over one thousand dollars per person!

Rabies isn't the only threat from handling raccoons, either. Enter raccoon roundworm *(Baylisascaris cyonis),* which can pose grave risks to human health. Although this parasite hasn't previously received much publicity, it is now becoming more widely recognized as a potentially serious problem.

This roundworm, common and widespread in raccoons, is especially prevalent in the Northeast and Midwest. It causes a disease, known as *larva migrans,* in over fifty species of animals, including humans. Larva migrans can cause blindness and is potentially fatal. Children are especially susceptible to this disease.

The eggs of raccoon roundworm are carried in the coon's feces and can live in the soil for years. To make matters worse, raccoons carrying this parasite don't display any symptoms of disease, thus making seemingly healthy "pet" raccoons a double hazard. Other animals infected with larva migrans exhibit symptoms somewhat similar to those of rabies, and are frequently reported as rabies suspects.

Warnings about these twin hazards seem to fall on deaf ears. My own state of Vermont is currently in the midst of an epidemic of the raccoon strain of rabies. The media and the Health Department issue constant warnings to avoid raccoons and other potential carriers of rabies and roundworm. Despite this barrage of publicity, my file on raccoons contains numerous bulletins from the Vermont Health Department veterinarian, Dr. Robert Johnson, featuring pleas such as this: "We are swamped with cases involving people caring for baby raccoons. Rabies and raccoon roundworms are a major concern. Leave Wild Animals in the Wild." It appears that so-called common sense is a most uncommon commodity nowadays.

Because of rabies and roundworm, if for no other reason, never, NEVER handle raccoons, including orphaned babies wandering about! Instead, call a qualified wildlife rehabilitator. Most state wildlife agencies have a list of licensed rehabilitators who know how to handle wildlife safely without undue risk to themselves.

The raccoon's scientific name comes from the Greek: *Procyon* descends from *pro,* before, and *kyon,* dog, while *lotor* means one that washes. Folklore has it that raccoons are very clean animals that wash their food before eating it, and it's true that when water is handy, raccoons will often dip food in it—hence the scientific name of *lotor.*

No one is quite sure why raccoons behave in this fashion, although it's certainly not a sign of cleanliness. It's been postulated that raccoons sometimes dunk their food because they lack salivary glands and need to moisten the food for easier swallowing. That theory is incorrect, since raccoons have normal salivary glands, but, saliva or not, perhaps wetting down comestibles renders them easier for the animal to devour. This raccoon trait remains something of a mystery, but one fact is abundantly clear: raccoons eagerly consume large quantities of food far from water without any scruples whatsoever. That should abolish any notions about raccoon cleanliness.

The name *raccoon* derives from the Algonquian *arakunem,* meaning "hand-scratcher," or "one that scratches with its hands." Early European settlers soon corrupted this into *raccoon,* and raccoon it has been ever since, but the animals are often simply referred to as coons, especially in rural areas.

Native Americans were keen students of the habits and abilities of the various wildlife species with which they had contact. This should come as no surprise; after all, animals were sources of food, clothing, tools, and other necessities. It's interesting to note, then, that the multitude of names given the raccoon by numerous Native American tribes nearly all make reference to its front paws. These generally translate into terms such as "touch things," "pick things up," "grasper," "handle things," and so on.

These names pay tribute to the raccoon's clever front paws, which are indeed almost as useful as hands. Those nimble little fingers are extraordinarily dexterous, and the coon's ability to pry, twist, lift, turn, push, jiggle, and otherwise gain entrance into containers and various other places where it shouldn't be is nearly legendary!

This manual dexterity is coupled with a great deal of apparent intelligence. Intelligence is a difficult thing to measure; witness the huge disagreements among experts over what constitutes human intelligence, let alone intelligence in other species. Still, most animal experts would concede that coons rate rather high on the intelligence scale. Put together the cleverness to devise ways to get into things with the dexterity to accomplish them, and you have the combination that can make the raccoon so destructive in many situations.

On cursory inspection, the raccoon's appearance might suggest at least a passing relationship to dogs, foxes, coyotes, and other members of the dog tribe. Indeed, the genus name *Procyon*—"before dog"—makes a bow in that direction. Raccoons are more closely related to bears than to dogs, however.

Nate, a wonderful black Labrador that we once had, certainly recognized that raccoons weren't related to dogs. Whenever he saw a fox or another dog, Nate's body and tail language simply radiated bonhomie. On the other hand, he was utterly infuriated by raccoons and missed no opportunity to kill one whenever he found it in our barn. Nate had many epic battles with this sworn enemy and—remarkably, since coons are savage fighters—never incurred a single wound from these encounters.

Although they're distantly related to bears, raccoons share a separate family with their nearest kin—ringtails and coatis—and hence aren't closely related to any other group of mammals. The ringtail *(Bassariscus astutus)* ranges from southern coastal Oregon through California and the Southwest down into Mexico. Marked in similar fashion to a raccoon, it looks much like a coon that's been stretched so that it will fit through a narrow tube. The larger coati *(Nasua narica)* has a long, slender tail with rings far less distinct than the coon's, as well as a less distinct facial mask; north of Mexico, it inhabits only the very southern portions of Arizona, New Mexico, and Texas.

Raccoons are classified as carnivores and are certainly meat-eaters, but in fact they are omnivorous. Fruit, acorns and other nuts, frogs, crayfish, salamanders, mussels, birds' eggs and young birds, insects, worms, carrion, and garbage—to say nothing of the contents of bird feeders, such as suet and sunflower seeds—are among the host of items that coons consume avidly. Then, of course, there is corn, particularly sweet corn. Raccoons are justly famous for raiding corn patches, and are the bane of many a gardener's existence. A family of raccoons can utterly devastate a large patch of corn in just one night, destroying many times what they actually eat. They know when the corn is just ripening, too; they'll pass by it night after night without paying the slightest attention until the corn is almost ready for humans to pick—then virtually annihilate the entire patch in a single evening of unbridled gluttony.

In balance with its natural habitat, as it once was, the raccoon was simply a normal predator—a valuable part of the ecosystem. Because of their extreme adaptability, omnivorous appetites, and cleverness, however, they've learned to take advantage of human activities. In particular, they seem to regard urban and suburban areas as their rightful domain; there they inhabit attics, crawlspaces, outbuildings, and various other man-made structures, meanwhile growing fat on garbage, to say nothing of the birdseed and suet so thoughtfully provided by humans. The result is an enormous overpopulation that spills out into rural areas and wreaks havoc on many creatures, including songbirds, ducks and geese, and frogs—to say nothing of the contents of human gardens.

Our daughter-in-law's uncle owns a building and home-repair business, and not long ago he had an experience that vividly demonstrates the scope of the problem. One of Albert's clients owned a home which had been vacant for some months, and called to say that there was an apparent leak in the roof; one of those ceiling lights with a concave glass disk below it was full of yellow liquid.

When Albert investigated, he found the attic full of raccoons—thirty-five, to be precise! The light fixture was full of urine that had dripped through the ceiling, and the attic was inches deep in feces. Now consider that for every instance such as this, there are dozens, and perhaps hundreds, of attics, crawlspaces, and outbuildings that harbor one, two, or perhaps as many as five or six coons. The total raccoon population of even a small city or suburb suddenly becomes staggering when viewed in this light.

Raccoons are excellent climbers and, aided by those facile front paws, their depredation on many birds has become a serious matter. One biologist told me that he has to be very careful about checking bluebird and other nesting boxes: raccoons have learned that following human tracks will often lead them to nesting boxes, which they can then rob. Unless some sort of barrier is placed around a tree or pole that holds a nesting box, a coon will climb to the box, reach in through the entry hole, and extract eggs and baby birds.

Further evidence comes from a study performed in an experimental forest in New Hampshire some years ago. Biologists located nests of various species of songbirds and then set up cameras with tripwires and flashes to catch nighttime nest robbers in the act. To no one's great surprise, the photos showed that by far the most destructive nest robber was the raccoon.

Raccoons are extremely vocal animals when they wish to be, and exhibit a wide variety of calls. These include a sound variously described as twittering or chirring, although it strikes me as more of a rattling; snarls and growls; and shrill, angry squalling that greatly resembles a fight between two tomcats.

Then there is another raccoon sound that few seem to have heard, but it's one I can verify from personal experience. One summer night a number of years ago, with the bedroom windows open, my wife and I were awakened by an eerie sound that seemed to come from some indeterminate place and distance. It's nearly impossible to describe the sound in writing, but it seemed like something from a movie about outer space—an *oinnnnggg, oinnnnggg, oinnnnggg, oinnnnggg, oinnnnggg,* with each *oinnnnggg* in a series pitched a little lower than the preceding one. Another way of describing the sound might be that it resembled the plucking of a string with a lengthy reverberation, or perhaps something produced by a synthesizer.

So weird was the sound, which simultaneously seemed both far away and near at hand, that the hairs on my neck rose, and I was covered with goose bumps. I don't mind admitting that I was frightened and began to wonder if there might indeed be some truth in tales of space aliens and flying saucers!

Finally, summoning what little courage I still retained, I took a flashlight and went out onto the lawn. A sweep of the flashlight at first revealed nothing, but then the sound suddenly seemed to come from a big maple tree by the corner of the house. When the beam of light scanned the maple, the culprits were revealed. Everywhere I shone the light, there seemed to be a pair of glowing eyes, and it was soon apparent that a mother raccoon and her large brood of little ones were in the tree and had been making that uncanny sound. I have no idea whether it was the mother, the young, or both making such an unearthly noise, but it clearly was the product of a raccoon's vocal apparatus.

Several years later I wrote a short item about this experience for the newsletter of a conservation organization. In response, I received a letter from a gentleman who told me that he had encountered the same strange noise and had been able to identify raccoons as the source. Evidently this is not a common raccoon vocalization, or more people would have heard it, but coons are most definitely capable of uttering such a sound.

Raccoons are prolific, which is one of the reasons for their present overpopulation. Usually four or five young, though sometimes as many as seven, are born in April or May, following a gestation of a little over two months. In forested areas, where there are no den sites conveniently provided by humans and their activities, raccoons prefer to den in hollow trees. In the absence of a hollow tree, however, a hollow log, a crevice among rocks, or even a burrow in the ground will suffice. The male coon, called a boar, plays no role in raising the young; indeed, a female will drive a male away from her den if he approaches too closely.

Raccoon kits are blind at birth; their eyes open in about three weeks, although they don't have full vision for a bit longer. They're able to run and climb in less than two months, and can accompany their mother on her mainly nocturnal travels after about ten weeks.

Raccoons are medium-sized animals. The total length of an adult, including an eight-to-ten-inch tail, runs from about two feet to a bit more than three feet. Weight usually varies from ten to thirty pounds, although a very large male may occasionally reach thirty-five pounds. No doubt in some situations, raccoons grossly overfed on garbage, birdseed, suet, and similar goodies will even exceed that weight.

As previously noted, adult raccoons are savage fighters that few predators want to tackle. Wolves and cougars were probably their principal enemies historically, but these big predators are now absent from most of the coons' range. Bears and bobcats may kill one occasionally, but humans are their main predators nowadays.

If adults are immune to most predation, raccoon kits are another matter, at least until they're well on their way to adulthood. When they first start traveling about with their mother, the kits can easily fall prey to great horned owls, large hawks, coyotes, fishers, bobcats, bears, and any other predator large and strong enough to tackle fairly small prey. By fall, though, they're able to defend themselves from the majority of predators.

Raccoons seem to enjoy water, especially wading around in the shallows searching for food. The mud along almost any stream or pond is apt to reveal quantities of little handlike prints—the evidence of this creature's predilection for wet areas. Coons are also very strong swimmers when the need arises, although they aren't particularly swift in the water.

With the approach of winter, raccoons put on layers of fat, much like miniature bears, and then go into winter quarters when really cold weather and/or deep snow arrive. But raccoons are not hibernators. That is, they don't go into a deep sleep like the woodchuck, or have special adaptations like the bear, to help them survive the long winter. Instead, they simply remain dormant during the worst weather, but emerge to wander about seeking food during warm spells. Thus raccoon tracks can sometimes be seen in the snow during a winter thaw.

At one time, winter dens were most often in hollow trees, although hollow logs, holes in the rocks, or burrows sometimes sufficed. As already noted, however, all that has changed. Now a high percentage of raccoons den in or under some man-made structure. Male raccoons mostly den in solitary fashion, but a mother and her nearly grown kits will often den together.

In reasonable numbers, raccoons are fascinating animals, well worth studying and observing. Unfortunately, they've now become a serious problem for humans and most especially for some species of wildlife. With wolves and cougars gone from most of the eastern United States and Canada, humans now exert the only real control on raccoon numbers, other than such nasty diseases as rabies and distemper. Given present conditions, including the unduly sentimental view that many have of the raccoon, we can expect their overabundance to continue.

16

Not a Cat, But . . . : The Red Fox

MYTHS

- Red foxes weigh twenty-five pounds or more.
- Red foxes roll, tumble, and leap in order to attract curious prey within pouncing distance.
- Red foxes are nonnative, brought here by European settlers.

LIKE A WIND-DRIVEN TONGUE OF FLAME, THE FOX FLICKERS THROUGH THE UNDERBRUSH OR, LIKE A CREEPING GROUND FIRE, MOVES SLOWLY AND STEALTHILY THROUGH THE GRASS IN SEARCH OF PREY. With its bright reddish-yellow or reddish-orange coat, neat black "stockings," alert ears with black backs, and white beneath its body and pointed muzzle, the red fox *(Vulpes fulva)* is unquestionably one of nature's loveliest and most elegant creatures. Even its scientific name, taken verbatim from Latin (*Vulpes,* fox, and *fulva,* reddish yellow) pays tribute to the predominant color of the red fox's handsome coat.

Red foxes aren't always red, although that's their usual color. Two noteworthy variations are the silver and cross foxes. The silver fox is basically a black phase of the red fox, so named because silver-tipped guard hairs grow out through the black in winter, giving the fox a silvery appearance. The cross fox is named for the dark fur, in the shape of a cross, spanning the shoulders and running down the back. These two variations, found mostly in far northern climes, were rare enough to be extremely valuable for their fur in the days before fur farms began producing them.

No account of the red fox's appearance would be complete without mention of its glorious tail, commonly called its "brush." This splendid

Red fox

appendage is so thickly furred that it has real substance if squeezed, yet it weighs virtually nothing. The tip of the brush is white and is known as the "tag." No other North American canid, including other species of foxes, has this white tag. Resplendent in its winter coat, a running red fox, magnificent brush seemingly floating behind it, is truly a sight to be treasured!

The fox's tail evidently has considerable value to its owner. In extremely cold weather, foxes often curl up tightly and cover nose, paws, and legs with that thick, luxurious brush; no doubt this helps to reduce heat loss substantially. Further, the tail is useful in fox body language to convey such messages as dominance, submissiveness, aggression, or playfulness.

The red fox is unquestionably a member of the dog family. Even a cursory glance reveals that it looks very much like a small, slender dog that bears little resemblance to the cat family. Yet biologist J. David Henry, author of a number of articles and books on the red fox, has demonstrated after years of research that red foxes have a number of very catlike—and hence quite undoglike—traits.

Henry, whose lengthy scientific inquiries have done for foxes what Maurice Hornocker has done for cougars and L. David Mech for wolves, has compiled impressive findings that document both catlike habits and catlike adaptations in the red fox. Consider hunting behavior, for example.

Red foxes hunt in much the same manner as small cats, and feed on the same prey. Whereas other members of the dog family tend to be endurance hunters that will often chase prey for long distances before exhausting it and finally killing it, red foxes are stealth hunters. Whenever possible, they stalk their prey and either pounce on it in very feline fashion, or try to run it down in a quick burst of speed. Further, just like a cat, they'll sometimes toy with small prey before finally dispatching it.

Where other members of the dog family hunt in groups, at least part of the time, red foxes, like cats, are solitary hunters. Even a mated pair of foxes usually splits up in order to hunt in lone fashion. This solitary behavior is perfectly suited to the red fox's stalk-and-pounce mode of hunting.

The fox has a number of physical adaptations that make it much more catlike than other canids (members of the dog family). One of the most striking is its eye structure, which is far more catlike than doglike. The red fox has two important optical characteristics in common with cats: vertical pupils that can narrow to a slit, thus dramatically reducing the light reaching the eye in bright sunlight; and a reflective membrane at the back of the eye, which causes light to pass over the retina twice, thus greatly improving night vision. Taken together, these two adaptations give the fox, like the cat, outstanding vision under light conditions ranging from deep darkness to brightest sunlight.

As almost everyone knows, cats can retract their claws, and, to a degree, so can red foxes. Partially retractable claws may serve at least two purposes. First, they permit quieter stalking, especially on stones or hard ground; second, they stay sharper, and sharp claws are an aid to both fox and cat in pinning small prey.

When a fox dispatches its prey, it does so in a manner similar to that of a cat. Most canids seize their prey and shake it vigorously from side to side as a means of killing it or rendering it helpless. Red foxes, on the other hand, have long, slender canine teeth, more like those of cats than of dogs. Using those long, sharp canines, the fox eschews shaking its prey and, cat-fashion, simply bites down hard to kill it.

The fox's vibrissae—what we usually call whiskers—are catlike, too. Where other canids have proportionately shorter whiskers, the fox sports longer ones in the manner of a cat. Even a fox's threat display toward one of its own species is similar to a cat's: its hair stands up, its back arches, and it turns broadside in a stiff-legged prance toward its rival. Anyone who's had much experience with house cats has seen this sort of performance, which is notably absent in dogs.

Red foxes prey on almost any creature they can catch and kill. These include large insects, such as grasshoppers and crickets; small birds; occasional larger birds such as grouse, pheasants, and ducks; squirrels; hares and rabbits; and snakes and lizards. They also eat berries and fruit when they're available, and the Song of Solomon proclaims, "Take us the foxes, the little foxes that spoil the vines." Evidently the fable of the fox and the grapes isn't entirely fanciful.

One other item—young woodchucks—can be added to this smorgasbord of fox treats. Several years ago, close to the woods that border our fields, a fox trotted past me only a few feet away, a defunct, half-grown woodchuck clamped securely in its mouth. The fox's entire demeanor, from head held high and forward to its high-stepping gait, immediately reminded me of the many house cats that I've seen carrying chipmunks and mice in that fashion.

Despite such wide-ranging tastes in food, red foxes are, above all else, preeminent mousers (as used here, "mouse" includes a variety of small, somewhat mouselike rodents, such as voles). Watching a mousing fox is a marvelous sight, one of the most arresting in the natural world.

We are indebted to J. David Henry for his insight into the various methods used by foxes to hunt different types of prey. He has observed, for example, that foxes hunting large insects, which are quite easy to catch, behave in a rather offhand, relaxed manner, but their demeanor when mousing is quite the opposite. Ears pricked up and head held high, the mousing fox is the epitome of alertness.

The fox has wonderful hearing, and, by turning its head slightly from side to side, can locate a mouse rustling in the grass with considerable accuracy. Using sight, smell, and hearing—but particularly hearing—the fox stalks forward step by cautious step. Finally, when it deems its prey within range, it crouches, sets its feet, and leaps in a high, spectacular arc, coming down with its extended front legs together, feet aimed at its prey in an effort to pin the hapless rodent to the ground. If successful, the fox either crunches down with its sharp teeth to kill the rodent, which it then gulps down, or carries its prey to a spot where it can play with it before administering the coup de grâce; in either event, this is very catlike behavior.

Although foxes are capable of leaping fifteen feet or more with considerable accuracy, most of their hunting jumps are much shorter—three to six feet. In a way, these shorter leaps are more spectacular than the longer ones, because the fox moves in a higher, narrower arc, ascending almost vertically and then plunging sharply downward in the same fashion.

The first time I observed a mousing fox, I initially didn't know what to make of it. The creature was on the far side of a field from me, and periodically it jumped high in the air to come down front feet first. The first thought that crossed my mind was that it might be rabid, although it seemed rather too vigorous for a rabid animal. Then, after several unsuccessful attempts, the fox came up with a mouse in its mouth, and the light suddenly dawned!

More recently, I observed a fascinating variation of this behavior. It was winter, and we had had a light rain that formed a rather hard, somewhat translucent crust of ice. A mousing fox that could either see or hear a rodent beneath the crust was performing its normal mousing leap over and over in a series of unsuccessful attempts to break through the crust. Periodically it moved about, no doubt in search of alternate prey—then repeated the futile series of jumps. I finally had to leave and never did learn whether the fox ever succeeded in its ill-starred quest.

Even if this fox was unsuccessful for two or three days because of the icy crust, it still had a good chance of surviving, because, when hunting is good, foxes cache surplus food by burying it. Unlike some mammals, red foxes cache each mouthful of surplus food in a different location, and these caches are often widely scattered. Extensive research has revealed that foxes retrieve and eat most of this cached food, finding each cache mainly by an extraordinary memory for its exact location. The fox's keen sense of smell also helps it locate caches, especially if the fox recalls only the approximate location of the cache, or if the site is covered by several inches of snow.

Foxes are also quick to learn where goodies come from, and some soon learn to follow a mower at haying time in order to snap up mice and voles killed or injured by the mower or deprived of the thick grass cover that makes them difficult to hunt. In fact, I've had some wonderful personal experiences in this regard, looking behind to find a fox trailing the haybine at what it regarded as a safe distance and gobbling up mice. The memory of one young fox, in particular, stands out. Following the haymower quite closely, it gorged itself on mice and voles until it could hold no more. Then it simply curled up like a tired child sated with goodies in a candy shop and fell asleep in the mowed field in full view of all!

Red foxes also exhibit at least two other types of hunting behavior. When a fox sights a small bird or tree squirrel on the ground, it employs a technique quite different from that used in mousing. Crouching very low, the fox stalks closer, apparently less concerned with being totally silent than with remaining invisible to the prey as long as possible. As it draws nearer to its prey, the fox speeds up its approach until it's actually running in the crouched position, then pounces at its intended victim and attempts to seize it in its mouth. More often than not, of course, the bird takes wing or the squirrel darts up a tree before the fox can capture it.

Yet another strategy comes into play when the fox hunts rabbits and hares. First it crouches down to stalk as close as possible, and then, when the prey sights the fox and flees, the fox attempts to run it down. The rabbit or hare takes evasive action, and the high-speed chase continues until the fox either gives up the pursuit or succeeds in bringing down its quarry.

One stratagem purportedly used by foxes is called "charming." According to oft-repeated accounts of this behavior, a fox sights potential prey—a rabbit or duck, for example—and proceeds to perform a series of antics to entice the curious prey close enough that the fox can pounce on it. These antics include rolling, tumbling, leaping, and chasing the tail, and reputedly mesmerize or "charm" the prey—hence the term.

This seems a dubious proposition at best. The red fox responds to prey instinctively, using one of the techniques already discussed—not by performing a series of antics. Further, prey such as rabbits normally shy away from movement and from the unknown, rather than being hypnotized by it.

It's also interesting that Henry, who has observed and analyzed the hunting behavior of large numbers of red foxes over the years, makes no mention of this behavior.

That isn't to say this type of incident has never occurred, however. As with many other mammals, foxes can indulge in what appears to be playlike behavior, running about, leaping, rolling, and tumbling. It's possible that under such circumstances a particularly curious and unwary individual of a prey species might draw closer and closer to see what was going on. If a hungry fox spotted the prey at that point, it would likely pounce on it. Whether such reported incidents are authentic is debatable, but fox behavior under such circumstances is almost certainly not a planned artifice for capturing a meal.

For a long time there was a debate over the origin of the red fox in North America, and many believed that the red fox was not native here. It was known that English settlers, anxious to continue their pastime of hunting foxes on horseback, had imported red foxes from England during the early period of European colonization. According to this theory, those imported red foxes multiplied and spread, as the continent continued to be colonized by Europeans. Red fox fossils have, however, been found in North America that date to well before the arrival of Columbus, thereby firmly establishing this creature as native.

Ask a number of people how much they think an adult red fox weighs, and most will give answers in the range of thirty pounds or more. This is a gross overestimation (a common phenomenon in assessing wildlife size). A really big fox will weigh only about fifteen pounds, and ten to twelve pounds is more usual. The fox's beautiful coat, relatively long legs, and magnificent brush combine to give an impression of much greater size, but a very small body resides within that handsome package.

Foxes can sometimes exhibit a playful nature. Golfers at one of our local courses still remember the pair of foxes that raised its litter in a den near a fairway. Whenever a golfer's drive landed in the appropriate area, a fox would trot out of the brush, pick up the ball, and retire to its den. There doesn't appear to be any biological reason for this behavior, other than pure enjoyment. Even if a fox mistook a golf ball for an egg the first time it encountered one, it would quickly learn better and abandon the practice. In this instance, the foxes continued to snatch golf balls throughout the summer. Regardless of the foxes' motives, the golfers were inevitably too enchanted by the sight to mind the loss of a ball! I was also recently told of a family of foxes that behaved in similar fashion at another golf course. This trait may be analogous to similar behavior by crows and ravens, as noted in chapter 13.

Although red foxes are very catlike in the many ways already outlined, they have other characteristics besides general appearance which tie them closely to their dog family relatives. This is especially true in their mating habits and family life, where their canid characteristics are most prominently displayed.

About late January, red foxes begin to seek their mates. It's easy to tell when the mating season has begun, at least in northern climes where there's snow on the ground. Simply follow a fox track to a place where the fox has urinated; there, a very strong, almost skunklike odor—absent at other times—proclaims that mating is in session. This smell is pungent enough to be detected at some distance if the wind is in the right direction, and, once experienced a time or two, is almost unmistakable.

Fox mating involves a substantial amount of foreplay, some of it quite spectacular. After considerable playful chasing in a sort of fox version of tag, the male (called a dog fox) and the female (known as a vixen) stand erect on their hind legs and put their forepaws against each other, almost like dance partners. Then they may again race after each other and repeat the performance. This courtship behavior can continue for most of two weeks, until it finally culminates in the actual mating. Then the two temporarily split up to again become solitary hunters for a time. The dog fox will then rejoin the vixen, ready for parenting duties, before the young are born.

As the birth of the young approaches, the pair begins to prepare a natal den. Usually this is in soft, well-drained soil where the digging is easy, and often it's an enlargement of a preexisting den dug by a woodchuck, a skunk, or even another fox. It may be in a pasture, a meadow, an abandoned field, or the edge of the woods, but it usually has some open space around it. The vixen commonly excavates several entrances, as well as various tunnels and chambers. An older vixen may have several natal dens within her territory, selecting one or another each year according to her mood.

The natal den, incidentally, is quite different from a much simpler type of den used for resting or escape from danger. Foxes usually have a number of these dens, which consist of a few feet of tunnel with a small chamber at the end. Here a fox can wait out a particularly bad spell of weather, such as a blizzard, or escape a variety of predators.

After a gestation of about fifty-two days, the vixen gives birth to a litter of three to six young, although larger litters sometimes occur. The little foxes are usually called kits, in another nod toward the fox's catlike characteristics, though they're also sometimes called pups or cubs. The vixen remains in the den for the first few days, nursing her kits. During this time her mate brings her food on a regular basis; this behavior is typical of the dog family, but not of cats. As the kits continue to grow and are gradually weaned, both parents hunt and bring them food.

If the den is disturbed at any time, or the parents feel that danger threatens, they'll quickly abandon that den and move their kits to an alternate site, as I once had an opportunity to observe. I was mowing our lower field and suddenly became conscious of two adult foxes that alternately appeared, disappeared, then reappeared again. At first I simply attributed their behavior to the habit of following the mower to pick up mice. After a time, however, I noticed that they seemed quite agitated, though I had no idea why.

It wasn't until we had baled the hay that I discovered the problem: the foxes had enlarged an old woodchuck hole in the middle of the field, and were clearly upset by our haying activities. Although I spent considerable time in concealment, watching the den, I never saw the foxes near it again, and it became obvious that they had abandoned it almost as soon as I began mowing nearby.

Initially the kits have a grayish coat, but after about five weeks it changes to a rather dull yellowish brown that usually blends well with the sandy soil around the mouth of the den. Around this time the kits also begin to eat small prey brought to the den by the parents.

Well before this, at less than four weeks, the kits begin to fight, often quite fiercely. This behavior continues for a number of days and isn't, as many people think, mere playful roughhousing. Rather, it's the means of establishing a dominance hierarchy, which has a great deal to do with which kits are fed first by the parents. If food happens to be particularly scarce, only the more dominant kits may survive, so this juvenile fighting is serious business.

As the summer wears on, the kits finally lose their brownish-tan color and are attired in the gorgeous orange-red coat which helps make the red fox, in the opinion of many, our most beautiful mammal. By this time the kits also begin to accompany their parents on hunting forays, and gradually start to forage a bit around the den in their parents' absence.

When autumn arrives, the young foxes—no longer kits—are ready and able to fend for themselves. Then they disperse and seek new territories. No one has determined why young foxes disperse in the fall, but the young males depart first and travel the farthest in seeking new territory; this, incidentally, seems typical of many young mammals, and may be a way of preventing inbreeding.

If these dispersing young encounter other foxes with established territories, they're apt to be in deep trouble. Foxes are highly territorial, and the resident dog fox or vixen will attack the intruder savagely and without warning. Though the youngster may flee, it may be caught and bitten one or more times before it manages to escape from the forbidden territory.

At some point a young male and female will enter a vacant territory and will locate each other. In this fashion another pair of foxes is joined, and the unending cycle that perpetuates the species continues.

In the days when free-ranging poultry were common, foxes had a bad reputation for snatching domestic fowl, as witness the delightful folk song that begins,

> *The fox went out on a chilly night, and prayed for the moon to give him light.*
> *He had many a mile to go that night, before he reached the town-o.*
> *He ran till he came to a great big pen, the ducks and the geese were kept therein.*
> *"A couple of you will grease my chin, before I leave this town-o."*

Although such depredations were no doubt exaggerated at times, thereby obscuring the fox's invaluable role in rodent control, some foxes unquestionably found ducks, geese, and hens easy pickings and took a heavy toll now and then on a farmer's flock. Indeed, my earliest recollection of wildlife involved a fox that came visiting, evidently attracted by our flock of hens.

My mother hurried into my bedroom to tell me that there was a fox by the henhouse, and that my grandfather was going to shoot it—not unreasonably, since the fox's proximity to the hens wasn't a good omen for their future survival! We raced to the window, and I had a moment to view the handsome creature. Then there came the loud report of Grandpa's gun. I must have blinked, and, presto! the fox vanished so swiftly that I never saw it go. Grandpa had obviously missed, but the fox clearly received the message, for it never returned.

A few years ago we experienced a far different sort of incident involving a fox and our domestic birds. At that time we had kept a small flock of free-ranging geese for several years, and our local foxes never bothered them. It was midwinter, with deep, soft snow, at the time of this event. We were awakened in the night by an ungodly honking and squawking from the little three-sided goose house just across the road.

I was in the midst of a nasty bout with the flu at that time, but I dragged myself out of bed the next morning to see what had happened. There was a dead and partly eaten goose, but the others were all missing. It turned out that they had flown to escape the danger and were here and there, floundering in the deep snow and unable to fly or make much progress by walking. They weren't the only ones that floundered, either. Feeling sick and miserable, I had to wade in the deep snow to retrieve them, one by one. The tracks showed beyond a doubt that the culprit was a fox, and my opinion of foxes temporarily dropped to an all-time low.

About noon the following day, I heard another uproar from the goose house and ran to the window. There was a pathetically skinny, bedraggled fox chasing a goose across the road and through our open barn door. I ran for my deer rifle, and as I reached the porch, the fox emerged from the barn *sans* goose. One shot put an end to the wretched creature's misery. It was in the last stages of mange, which was obviously going to be fatal within a short while, so I was happy to be able to end its suffering. That was the only time a fox ever bothered our geese, and no doubt this fox, too weak to catch its normal prey, went after them out of desperation.

Mange is by no means the only disease to afflict foxes. At times, rabies sweeps through fox populations, and in fact one of the most common strains of rabies is known as the fox strain. For this reason, humans should avoid contact with foxes, and should especially resist the temptation to rescue orphaned baby foxes. Any fox that acts in a suspicious fashion—excessively tame, aggressive, porcupine quills in its mouth, drooling, wandering about aimlessly, or other unusual behavior—should be strictly avoided and reported immediately to the proper authorities. If contact does occur with a fox, this should promptly be reported to health authorities so that the animal can, if possible, be tested for rabies, and the need for rabies shots for the person or persons involved be assessed by an expert.

Not all foxes that act sick have rabies, for the species is also susceptible to distemper. This can cause a fox to act exceedingly lethargic, drag its hindquarters, and exhibit other signs of illness. It should always be assumed, however, that such an animal is rabid, and contact with it should be carefully avoided.

There is one other aspect of red fox behavior—vocalization—that's worthy of discussion. Red foxes are credited with numerous different cries, calls, and sounds. Although they don't howl, and aren't as vocal as their larger relatives the coyotes and wolves, foxes are by no means silent. For instance, they have a high, shrill alarm bark that is often used to warn the kits of danger once they're old enough to leave the den.

Our own family has experienced only, at least to the best of our knowledge, one fox vocalization, and we learned about it in a most unusual way. For years my wife and I would be awakened at times by a strange, high-pitched sound. At a distance, it sounded almost like a deer snorting, although less breathy and more vocal. As it drew nearer, it sounded more vocal still; it's a difficult sound to describe, but is a sort of high-pitched *nyaaaaa, nyaaaaa.* We puzzled over this cry many times over the years, wondering what creature could possibly be making it.

Our black Labrador, Nate, gave every indication that he regarded foxes as little dogs that he'd dearly love to play with. Whenever we saw a fox out the window and summoned Nate, he'd wag his tail gently and whine longingly, then run back and forth between window and door, hoping to be let out—precisely the behavior that he exhibited when he saw another dog. Having had no direct experience with foxes, it never dawned on him that the "little dog" might not desire his companionship.

One day, accompanied by Nate, we were returning to the barn with a pickup load of baled hay from our lower field. There were a couple of electric fence gates that had to be opened first, and as we rounded a corner to open them, we suddenly heard that familiar and mysterious *nyaaaaa, nyaaaaa.* Looking up, we saw a gorgeous fox sitting in the middle of the farm road, perhaps forty yards away. Even as we spied the fox, we saw its mouth open and heard the unmistakable *nyaaaaa.*

I looked down at Nate. His eyes bulged, and he seemed momentarily paralyzed. Here was his longed-for playmate, just waiting for him. The moment beckoned. Shaking off his momentary astonishment, Nate raced toward the fox, which calmly sat there and uttered another *nyaaaaa.* Then, when Nate was about twenty feet away, the fox leaped up, turned on its afterburners, and departed in a flash of bright color, Nate in hot pursuit. The result was foreordained: Nate was fast, but he was no match for the speedy, buoyant fox. Some minutes passed before a crestfallen Nate, puffing like a steam locomotive, reappeared, the desire for the chase wrung out of him by all-out exertion and blighted affection.

This isn't the end of the story, however. That same night, and for several nights thereafter, the fox came onto our lawn just a few feet from our bedroom window and proceeded to say *nyaaaaa, nyaaaaa* a number of times before going about its business. This had never happened before and never happened again. Our conclusion was inescapable: with its keen sense of smell, the fox knew exactly who had chased him and where he lived, and he was taunting poor Nate, whom he clearly regarded as no threat at all!

Fox kits are vulnerable to a number of predators, such as hawks and owls. Adults, however, have few enemies other than man—and human predation is at a rather low level. The most serious obstacle faced by the red fox may be the spread of the coyote throughout the United States and much of Canada. Coyotes don't like foxes and will kill them whenever they can, although most of their predation is on fox kits. The number of foxes killed by coyotes is relatively small, though; the major damage is done by coyotes usurping what was formerly red fox territory and driving the foxes out.

Various studies have confirmed that coyotes guard their territory against foxes and force them out. This by no means forecasts the elimination of the fox, because the foxes manage to find territories in the interstices between coyote territories. Rather, it signifies a shift in the balance of predators, with the foxes replaced to some degree by coyotes.

Despite the pressure from coyotes, the red fox is adaptable enough to continue thriving in many areas. In some instances they've moved closer to human dwellings; apparently they've learned that in general the wary coyotes, despite exceptions in some heavily settled areas, are less apt to locate their territories in proximity to human activities. Since it has relatively few enemies and considerable adaptability, we can look forward to seeing this especially lovely and thoroughly atypical member of the dog tribe.

Coyote

17

Wile E. Indeed: The Coyote

MYTHS

- The eastern coyote is nearly as large as a wolf.
- Those animals out there are coydogs.
- Coyotes mainly hunt in packs.

MOST AMERICANS ARE FAMILIAR WITH ROADRUNNER CARTOONS IN WHICH THE HUNGRY COYOTE, WILE E. COYOTE BY NAME, IS FOREVER STYMIED BY THE CLEVER ROADRUNNER, AN ADVERSARY THAT IT SEEKS TO CONVERT TO THE COYOTE VERSION OF KENTUCKY FRIED CHICKEN. At every turn, the hapless coyote is battered, beaten, and outwitted, going hungry despite his best efforts.

The reality is quite the opposite, for the coyote *(Canis latrans)* is Wile E. indeed! It's a survivor *par excellence,* able to overcome all efforts to eliminate it and to thrive where most other animals could barely exist. This clever, wary, versatile predator, once regarded as primarily a creature of the West, has not only survived all attempts to exterminate it, but has blithely expanded its range all the way to the East Coast in recent years.

The coyote's name, incidentally, comes to us through Spanish from the Nahuatl word *coyotl.* (Nahuatl was the language of the Aztecs and certain other Indian groups.) Westerners usually pronounce its name *kai o ti,* while Easterners mostly opt for *kai ote.*

Coyotes are nearly as controversial as wolves, able to polarize opinions as few other creatures can. It's the feeding habits of coyotes that make them so controversial, bringing a deluge of both opprobrium and encomium upon their furry pates.

To many sheep growers and some human hunters, they're the devil incarnate, wanton, depraved killers, the destroyers of valuable sheep, small game, and deer. To others they're God's Dog (a term originally coined by Native Americans), a nearly sainted creature that kills only the old, the weak, and the sick, thereby maintaining the genetic health of the prey species. As a result, attitudes toward the coyote range from those wanting to kill none of the coyotes all of the time, through those wanting to kill some of the coyotes some of the time, to those wanting to kill all of the coyotes all of the time—preferably with tactical nuclear weapons!

Either extreme misses the point: coyotes are remarkably resilient and adaptable predators that may be viewed as either good or bad, depending on the situation. For example, coyote predation on sheep is no laughing matter. Coyotes devour as many as a quarter of a million lambs and sheep annually, thereby costing the sheep growers many millions of dollars. Some ranchers say that they lose up to a quarter of all their lambs to coyotes, and that coyote predation can put them out of business when prices of wool and meat are low.

Then there's the eastern coyote to consider—the newcomer on the block, so to speak. One of the most visible signs of the coyote's remarkable resiliency and adaptability has been its steady march eastward. Over the past few decades it has spread from its historic bastion in the West all the way to the Atlantic Coast. As a recognized subspecies, it has now become well established and common throughout most of the eastern United States and parts of eastern Canada.

The most noteworthy difference between the coyote of the West and its eastern brethren is size. Although the difference between the two isn't huge—on the order of about ten pounds—it's significant. The eastern subspecies of the coyote is unquestionably bigger and stronger than the various coyote subspecies in the West. (There have been nineteen subspecies listed in North America, although these fine subdivisions are apt to be in a state of flux.) According to most scientists, this size increase is due to the fact that the western coyote, during its eastward migration, picked up wolf genes by interbreeding with a small subspecies of the wolf in Canada.

Despite this very real difference in size, it's easy to exaggerate it. Consequently, eastern coyotes are often popularly regarded as "nearly as big as a wolf," though they're in fact nowhere near that large. True, occasional outsize specimens are found that are about as large as a small wolf. The Maine record is sixty-eight pounds, and an animal believed to be a coyote, weighing seventy-three pounds, was killed in Vermont. This specimen, however, has yet to be confirmed as a coyote by DNA analysis. In any event, very few specimens exceed fifty pounds, and most adult eastern coyotes fall into the thirty-to-forty-pound range.

Coyotes in the East are widely referred to as "coydogs," a misnomer grounded in a smidgen of truth. As coyotes slowly worked their way eastward, state by state, the coyote pioneers in many areas were few and widely scattered. Under those conditions, they sometimes had great difficulty in finding a mate, and eventually mated with a domestic dog. Thus there actually were a few coydogs as the leading edge of the coyote immigration made its way eastward. Coyotes much prefer to mate with other coyotes when given the opportunity, however, and as coyote populations increased, coyote/dog crosses virtually vanished.

Moreover, although the pups from a coyote/dog cross are technically fertile, they might as well not be, for all practical purposes. The coyote/dog cross results in a shift in the time of breeding, so that the coydog hybrids have their young in winter rather than spring. Coupled with this, the male coydog, unlike the male coyote, takes no part in raising the pups. This "double whammy" virtually guarantees that coydog offspring won't survive. The animals that inhabit the East are not coydogs, but are most emphatically eastern coyotes, a true-breeding subspecies of the coyote.

Eastern coyotes are also extremely controversial. For one thing, they too attack sheep, although sheep raising is far less prevalent in the East than in the West. Moreover, under some circumstances eastern coyotes can be significant predators of white-tailed deer, North America's premier big game animal. For a long time, most wildlife biologists believed that coyote predation on deer was generally limited to the spring fawn drop, and that evidence of deer in coyote scats at other times of the year was largely due to feeding on deer carrion. Lately, however, research by Maine wildlife biologist Gerald Lavigne has altered that opinion substantially.

Given two conditions—very deep snow and inadequate mature softwood cover in deer wintering areas—Lavigne found that coyote predation on deer can be very substantial. Still more interesting and significant, he discovered that coyotes under those conditions readily killed healthy deer in their prime—not merely the weak, sick, and old, as had previously been believed. Once a deer is floundering in deep snow, it apparently doesn't matter much whether it's healthy or infirm; in either case, coyotes can soon exhaust the deer and bring it down.

Although coyotes under these conditions can reduce the number of deer available to humans in limited areas, there's no evidence that they're decimating deer populations on a wide scale in the East, as often charged.

Indeed, the number of deer taken by human hunters in eastern states has risen substantially—and sometimes dramatically—during the very years when coyote populations were also growing and expanding. In many areas where coyotes are abundant, state wildlife agencies have steadily liberalized hunting regulations in order to hold deer numbers within biologically sustainable limits. Although the eastern coyote unquestionably preys on deer, it's evident that it isn't a limiting factor in whitetail numbers except under rather drastic winter conditions.

Undeniably, coyotes can and do cause major problems and substantial economic loss in several ways, but the equation is more complicated than that. Coyotes are great opportunists with catholic tastes. Their primary food sources aren't such things as sheep and deer. Rather, they feed heavily on mice, voles, ground squirrels, woodchucks, rabbits, carrion, frogs, snakes, lizards, house cats, small domestic dogs, garbage, large insects, seeds, berries, fruits, and other plant materials—in short, on almost every kind of food that's edible. Their penchant for eating fruit, incidentally, is also making the coyote *animal non grata:* in some areas where watermelons are grown commercially, coyotes have developed such an affinity for the juicy melons that growers are beginning to shoot them.

The difficulty is that it's relatively easy to quantify coyote damage to sheep, deer, and other creatures valued by humans, but extremely difficult to assess the value of coyote control of mice, voles, rats, woodchucks, and other creatures that also cause a variety of damage. Unquestionably, coyotes do a great deal of good in this regard, but that's small consolation to someone suffering serious coyote depredation on his or her flock of sheep.

This leads to the thorny and contentious issue of coyote control. Millions upon millions of dollars have been invested in coyote control programs, especially in the West. Coyotes are assailed in a variety of ways: by a division of the U.S. Department of Agriculture, by various state agencies, by coyote hunting contests, and by the efforts of individual ranchers. In fact, it's estimated that approximately 400,000 coyotes are killed each year by a variety of methods.

While the biggest coyote control programs are found in the West, the East has by no means been totally lacking in such efforts. As coyotes became well established in the East, there were loud cries to "do something" about this new menace. "Doing something" was usually vague, but often took the form of calls for bounties and other incentives to kill as many coyotes as possible. This attitude of "let's kill every one of the bastards" ignored mountains of evidence that such incentives were useless, and all too frequently led to expensive and ineffective efforts to control coyotes.

How effective have coyote control programs been, especially in the West? While they've temporarily reduced coyote populations in some localized areas, they've had little long-term effect on coyote populations. Indeed, there are those who argue that the control efforts have actually increased coyote populations because they've stimulated coyote reproduction. Ranchers counter that the control programs have either reduced coyote numbers in local areas or, at the very least, have prevented greater coyote numbers from causing even higher losses of sheep.

Before further exploration of the knotty problem of coyote control, let's examine the reasons for this animal's remarkable resilience. Why have coyotes been able to survive—and even increase their range and numbers—in the face of the most expensive and intensive animal control efforts in North America? The answer lies in three things: wariness, adaptability, and reproductive response.

Coyotes are known for their extreme wariness. One successful coyote trapper, for example, terms them the most difficult of all species to trap, even more so than the notoriously wary red fox. Likewise, coyote hunters can attest to the extreme cautiousness of the species. Although it's possible to hunt and trap them successfully, it's no easy matter.

Coyotes have also shown exceptional ability to adapt to new and changing habitats far different from their traditional ones. Once largely denizens of the prairies and deserts, they now inhabit densely forested areas, suburbs, and even parts of our largest cities, such as Los Angeles and New York!

Wariness and adaptability alone, however, probably wouldn't have saved the coyote from major reductions in its population, at least in many areas. Here's where the coyote's third line of defense comes into play: as soon as coyote numbers show any substantial decrease, the remaining coyotes begin to produce larger litters with a much higher rate of survival among their offspring. It's this trait that so effectively defeats the long-term success of coyote control programs. Even though temporary reductions may be achieved locally, the coyote population soon recovers.

Because of their depredations, real or imagined, on sheep and other livestock, game populations, pets, and other things valued by humans, coyotes generally elicit the question, "*What* should we do to control coyotes?" Others, perhaps a bit more thoughtful, ask, "*Should* we attempt to control coyotes?" Actually, those are the wrong questions; the proper one is "*Can* we control coyotes?" As far as long-term control over any sizable area is concerned, the answer is a resounding "No!"

Merely consider the facts dispassionately—admittedly a difficult task when they involve a creature that inspires such diametrically opposed and passionately held views. After many decades, vast efforts by hordes of people, and the expenditure of countless millions of dollars, there are more coyotes than ever. What's more, they've expanded their range enormously during that time, and now can be found almost everywhere.

I recall rather vividly the wise words of my friend Benjamin Day Jr., who at that time was chief of wildlife management for the Vermont Fish and Wildlife Department. Coyotes were just becoming widespread in Vermont, and there were loud cries that we should "do something" to control or, if possible, eradicate them. In response, animal-rights activists, who had adopted the Native American term of God's Dog, rushed to the animal's defense and worried publicly that this recent arrival to the East would be exterminated.

When I broached these conflicting notions to Ben, he suffered a near-fit of laughter. "Consider this," he told me. "In parts of the West where almost literally there's not a bush to hide behind, humans have tried everything to eradicate the coyote. They've tried guns, traps, dogs, gassing dens, hunting from airplanes, a variety of poisons, and anything else that could be devised—all without success. If they couldn't succeed under those conditions, how does anyone think that coyotes can be controlled, let alone eradicated, in a state that's 80 percent forested?"

Because of the general lack of success in controlling coyotes, there are those who advocate dropping all coyote control. They believe that if we stop killing coyotes, populations will actually decline, and depredation on sheep and other valuables will be reduced. This approach has its own problems, though. For one thing, it's an unproven theory that may or may not work. Second, even those advocating this approach admit that coyote numbers would probably increase for a time until the coyotes settled into a new and more natural equilibrium. Under those circumstances, agricultural losses would temporarily increase—something that sheep ranchers and others would be highly unlikely to tolerate.

Does all of this mean there's no course of action that promises greater success than our present efforts? No: a multifaceted strategy might work considerably better than either the present extreme of massive coyote killing or the alternative extreme of no coyote killing.

Because only a relatively small percentage of coyotes actually cause serious agricultural damage, targeting control efforts more precisely toward those specific animals would save a substantial amount of money. That money, in turn, would likely be far more efficiently used for several things. These include providing sheep growers with trained guard dogs or llamas (the latter reputedly hate coyotes and drive them away), and helping with the cost of better fencing. Certainly this more diverse strategy won't eliminate the problem of coyotes killing livestock, but it might well reduce it substantially. This approach isn't sloppy sentimentality about killing coyotes; rather, it's a practical view of value received for money invested.

Coyotes are justly famous for their vocal prowess—a fact recognized by the animal's scientific name. *Canis latrans* is Latin for "barking dog," and if barking is interpreted broadly to encompass a whole range of sounds, the name is certainly apt! Coyote calls range from an extremely doglike bark through high-pitched yips to full-fledged howls. The sound of a coyote family vocalizing at full volume on a dark night is both eerie and exciting!

All of this barking, yipping, and howling probably serves several purposes. It undoubtedly allows family members to stay in touch with each other, and serves as a sort of vocal territorial marking system. Also, at least in our area, coyotes tend to be most vocal in late summer and early autumn, when the pups are learning to hunt with their parents. It appears that when the family is either in hot pursuit of prey or has actually killed it, there's much excited yipping, yammering, and howling by the whole group. At other times, perhaps, coyotes simply howl for the fun of it, or as a social activity.

Western coyotes are sexually mature and begin breeding at one year of age, but the larger eastern coyote usually takes another year to reach breeding age. Mating, at least in northern climates, takes place in February. The pups are born after a gestation that averages sixty-three days. Litter size can be highly variable, from as few as two to as many as ten, though four to six is the norm. As already discussed, if the coyote population declines substantially, both litter size and survivability of the pups will increase.

With the onset of winter, most of the pups, now nearly mature, will disperse to seek their own territories, although they may rejoin their parents to cooperate in hunting big game. Still, by the time the following spring rolls around, most of the young have gone off on their own.

Coyote are often referred to as pack animals, but they aren't, at least not in the sense of the large, extended families that constitute wolf packs. The basic coyote social unit consists of a mated pair of adults plus their offspring less than a year old. Occasionally a yearling will remain behind to help its parents raise the next year's litter of pups. In the West, particularly, a handful of family members may augment the breeding pair and their offspring of the year, so that the family group can contain five to seven adults and perhaps a half-dozen pups.

On the other hand, coyotes can also operate singly and in pairs, so they aren't automatically tied to group behavior. Although the larger aggregations of perhaps a dozen coyotes can be considered a pack, they aren't typical, and a more usual situation is a family of four to seven or eight hunting together in late fall and early winter, then gradually diminishing until they're mostly hunting alone or in pairs.

Experiences with wildlife aren't always serious and inspiring events; indeed, they can be downright hilarious. A coyote was responsible for an event of the latter sort not long ago. I was deer hunting and had just entered the woods. As I moved cautiously over a little knoll, a movement to my right caught my eye. My first thought was that it was a deer, but it turned out to be a coyote headed in my direction.

I stood absolutely motionless and watched as the coyote trotted past me no more than twenty-five feet away. There was no wind that I could detect, and the coyote went by unconcernedly, without so much as a glance in my direction. Then a tiny breath of air must have carried my scent to the coyote, which exploded into action.

The ground was bare, and so frantic were the coyote's efforts to escape that, like a car spinning its wheels, its hind feet spun wildly in the leaves. Then, without warning, the coyote gained traction and shot forward like an arrow from a bow. As it accelerated, it threw a backward glance at me. This was a major mistake: a branch caught the rocketing animal under the chin as it turned its head forward, and the hapless creature was knocked half off its feet. Finally, the coyote managed to get its act together and depart with whatever speed and remaining dignity it could muster. For my part, I stood there shaking with laughter, for the whole performance was uncannily like something from a Roadrunner cartoon! So much for the notion that animals always perform efficiently.

Despite this temporary display of ineptness, the coyote is not a cartoon character. Yes, the Roadrunner will go on flattening Wile E. Coyote in the world of cartoons, but meanwhile the real coyote is here to stay. Whether one loves it, hates it, or takes a much more realistic view that this is simply a very interesting predator that is now part of our ecosystem, the coyote is an extraordinary survivor!

18

Bad Guy/Good Guy: The Timber Wolf

MYTHS

- Wolves are a threat to humans.
- Wolves are "good guys" that kill only the old, the weak, and the sick.
- Wolves are "bad guys"—cruel, wanton killers of everything in sight.
- Wolves are in danger of extinction.
- Wolves are nearly always gray.
- Wolves live mainly on mice in the summer.
- Wolves howl at the moon.
- Wolves always mate for life.

ALTHOUGH THE COYOTE IS HIGHLY CONTROVERSIAL, IT HAS TO PLAY SECOND FIDDLE IN THAT REGARD TO ITS BIG COUSIN, THE WOLF. Of all the animals on the North American continent, the timber wolf, also called the gray wolf *(Canis lupus),* has unquestionably been the most hated, feared, reviled, and relentlessly persecuted. Europeans who settled the continent regarded their new home as a howling wilderness, and sought to tame it by making it as much as possible like the "civilized" landscape that they had left behind. Predators, especially large ones, were regarded as a major component of that wilderness; accordingly, the colonists despised them and sought to eliminate them as an important part of wilderness taming. Still, they reserved a special brand of hatred for the wolf.

The reasons for this fear and hatred are complex and not entirely clear, but a major one was the attitude that the settlers brought with them from

Timber (gray) wolf

Europe. The wolf is cast as the Bad Guy in numerous fairy tales—"Little Red Riding Hood," "The Three Little Pigs," and "Peter and the Wolf"—and even in songs, such as "Who's Afraid of the Big Bad Wolf?"

Like most children of my generation, I was raised on such stories. One melodramatic tale in particular imprinted itself on my mind. It involved a huge, slavering pack of wolves relentlessly pursuing a sleigh and its human occupants across the snowy steppes of Russia. Aboard the sleigh were various supplies the family was bringing home (the frozen steppes in midwinter seem an odd time and place to be doing one's shopping, but that doesn't occur to a child).

Just as the wolves were about to attack, a side of bacon was dumped overboard, and the wolves stopped to devour and fight over it. They quickly resumed the pursuit, however, and soon were about to attack the horses and people again. Then the process of dumping food was repeated, and so on. Just after the last morsel of food had been jettisoned, and doom seemed imminent, the horses, with sleigh and occupants, dashed into their barn, barely ahead of the frenzied wolves!

Whether such sensational chase scenes ever happened in Europe is not known. There is speculation that European wolves might actually have attacked people for several reasons. Rabies was one, and starvation was another. Lacking the means to eliminate wolves directly, burgeoning European populations were constantly clearing more land and usurping wolf habitat. With habitat and natural prey largely gone, wolves no doubt turned to livestock as a food source—and possibly, under desperate circumstances, even to humans.

A third possible reason for wolf attacks has also been postulated. During the Middle Ages, bodies of some humans—executed criminals, outcasts, plague victims, paupers, and similarly unwanted characters—were often unceremoniously dumped without burial outside cities. Although wolves normally don't feed on carrion, they might conceivably have done so if nearly starved. In this manner, the theory goes, they might have lost an inherent fear of humans, and then turned to killing live humans.

We'll never know whether wolves actually killed people in Europe in those days, or whether these are merely monstrous exaggerations stemming from centuries of folklore, but let us dispense with one myth right away: North American wolves are no threat to human safety. There have been almost no documented attacks on a human by a non-rabid wolf in North America. Even the handful of attacks have been more or less provoked and involved mitigating circumstances. For instance, a wolf knocked down a hunter who was wearing camouflage and had covered himself with liberal applications of deer scent; when the wolf discovered its mistake, it promptly fled, probably more terrified than the hunter!

Wolves also enjoy—if that's the right term—an unsavory reputation as bloodthirsty beasts that kill for pleasure. In part, this image rests on their depredations on domestic livestock, which may strike the angry livestock owner as excessive and unjustified.

Wolves do indeed kill cattle and sheep—a trait noted from ancient times, as in the fable about the boy who cried wolf once too often, or in Lord Byron's memorable opening line of "The Destruction of Sennacherib": "The Assyrian came down like the wolf on the fold. . . ." Wolves, though, like other predators, kill only in order to live and not to satisfy some perverted blood-lust. To a wolf, a sheep is only a slow, dim-witted version of a small deer, and a cow nothing more than a physically challenged moose, elk, or bison.

In recent years, the North American timber wolf's image has undergone a major refurbishing in some quarters, though certainly not in others. Led by Farley Mowat's popular book *Never Cry Wolf* (more about that later), the wolf has become Mr. Good Guy to many, the persecuted, misunderstood keeper of sound genetic stock in caribou, moose, elk, deer, and other species, by selectively removing the weak, the unfit, the sickly, and the old. While this image is based on considerably more truth than are the old stereotypes of wolf as Bad Guy, it's often grossly overdone.

Conferring predatorial sainthood on the wolf isn't much more edifying than casting the animal in the role of unredeemed devil. The truth is simply that wolves are wolves and should be viewed on their own terms as one of the most efficient predators of large North American mammals. As such, they evolved as part of an ecosystem that predated humans on this continent. Ascribing all manner of human qualities to them—good or bad—is no help in understanding their place in nature's scheme, both historically and today.

Wolves are widely thought of as endangered. They aren't—at least as a species—although they're listed as endangered in the lower forty-eight states. (Minnesota wolves, with an estimated population of 2,500, are listed as threatened, while two reintroduced populations in the West are considered "experimental populations" under the Endangered Species Act.) Despite this endangered status in most of the United States, there are over fifty thousand wolves in Canada and over seven thousand in Alaska, so the total number of timber wolves in North America is approximately sixty thousand. Now that wolves are no longer relentlessly killed, the biggest threat to their long-term survival is habitat loss.

Canis lupus is Latin for "wolf dog" (*canis,* dog, and *lupus,* wolf), and we say that wolves belong to the dog family. Actually, this is backward; dogs are members of the wolf family. Domestic dogs have without question descended from wolves, although the timing and mechanism for this descent is not well understood.

Until recently, it was pretty much taken for granted that roughly fifteen thousand years ago, our Stone Age ancestors domesticated a few wolves, thereby creating the dog. One theory is that they might have raised wolf pups and tamed them. A second theory holds that some wolves, less fearful than the rest, hung around human encampments or villages, looking for easy pickings. From that beginning, these wolves gradually became tamer and tamer, increasing their tolerance of humans by breeding among themselves until at last they essentially brought about their own domestication. At that point, humans were able to complete the process, perhaps viewing these primitive dogs as partners in their own hunts.

Then, in 1997, University of California biologists announced their belief that dogs and wolves probably parted company more than 100,000 years ago. Their findings, based on DNA analysis of many breeds of domestic dogs, as well as of wolves and wild canids worldwide, are, rather predictably, highly controversial among scientists. No doubt it will take years more of study to sort out these theories and reach agreement on the most plausible one.

This interesting debate aside, no one disputes that domestic dogs descended from wolves at one time or another. Although it seems difficult to believe, all domestic dogs, from tiny Chihuahuas to huge Great Danes and Saint Bernards, owe their ancestry to wolves. Although these and most other varieties of domestic dogs bear only the most passing resemblance to their wild forefathers, there are still breeds, such as malamutes and German shepherds, that look very much like wolves.

Often called the gray wolf, timber wolves are commonly thought of as almost always being gray, and the majority of them are, but in the more northerly latitudes encompassed by Canada and Alaska, black wolves are common, and on Arctic islands wolves are white. All shades in between these colors, as well as blends of shades, are also found. It would be a mistake to think of "gray wolves" as exclusively gray.

How big are wolves? The largest ones, found in parts of Canada, Alaska, and Russia, may occasionally reach 175 pounds. However, most of the large northern timber wolves are much more likely to be in the range of seventy to 120 pounds. Farther south, wolves become progressively smaller, evidently obeying the same biological imperative that governs their prey species (see page 227). Female wolves are also about 20 percent smaller than males. Although a wolf of, say, ninety pounds may not sound like a very large animal, bear in mind that these are exceedingly lean, powerful animals!

Wolves are highly social animals, and the principal and all-important unit of wolf society is the pack. A pack is headed by a breeding pair, called the alpha male and alpha female; these terms come from *alpha,* the first letter of the Greek alphabet, and the alpha male and female are first—the leaders—among the pack members. The remainder of the pack consists of their offspring from two or more years and occasional older, nonbreeding relatives. Thus the pack is really an enlarged or extended family unit.

How large is a wolf pack? Packs in the North, where prey is larger, tend to be bigger than packs farther south, where prey is usually smaller, but this isn't always true. In general, wolf packs contain six to ten animals, but packs of twenty or more aren't rare and can number as high as about thirty. Still, it would be erroneous to think of wolf packs primarily in terms of these very large packs, which are definitely the exception.

There is a strict hierarchy within the pack. The alpha pair is indeed at the top, with every other wolf in a descending order from them. More or less jointly, the alpha male and female rule the pack. They lead the pack on hunts, are at the forefront of most kills, and feed first—or at least on the choicest parts—after the kill. They also bear the primary responsibility for raising the pups, aided by the subordinate members of the pack.

Below the alpha pair come the others at various levels in the hierarchy of the pack. At the very bottom are the omega wolves, so named from *omega,* the last letter in the Greek alphabet. Although life is often harsh and difficult for all wolves, it's doubly so for an omega wolf. That unhappy individual is sometimes so abused by all the other higher-ranking wolves that it's driven right out of the pack!

Wolves communicate in a wide variety of ways, of which howling is by far the best known and most impressive, and is the one for which wolves are properly noted. It's a special sound, for it contains something so wild and primitive that it evokes equally primitive responses in humans, such as hair rising on the back of the neck.

Wolves howl for a variety of reasons. One of these is to locate one another, for members of a wolf pack may inhabit huge territories in areas where big game is scarce and widely dispersed. Under such conditions, pack members may become widely separated, and howling helps to reassemble the pack.

Scientists have also observed that wolves howl as a group when excited or disturbed, particularly when they hear the howling of another pack. At such times, group howling may serve as a territorial safeguard—a warning to the other wolves that the territory is occupied and they enter it only at their peril. At other times, group howling seems to be a social activity accompanied by dominance displays by various pack members. On still other occasions, this sort of communal howling simply appears to be an enjoyable social activity, sometimes likened to a community sing.

Wolves are widely depicted as howling at the moon. However, despite the lines from Shakespeare's *Much Ado About Nothing,* "Now the hungry lion roars, / And the wolf behowls the moon," that's pure myth. Perhaps this old chestnut started when someone saw wolves pointing their noses skyward to howl on a moonlight night. No matter how many reasons wolves may have for howling, the moon most emphatically isn't one of them!

Wolves aren't limited to howling, though; like their dog descendents, they also make a variety of other sounds, including growling, snarling, barking, whining, and whimpering.

Though howling and other vocalizations are the most obvious means of wolf communication, they're far from the only ones. Scent marking by urination and defecation constitutes the second important method of territorial delineation. Urination, in particular, marks the pack's boundaries, and here the alpha male, and sometimes the alpha female, engage in the sort of raised-leg urination so familiar among domestic dogs. Presumably this raised-leg urination places urine at a higher level, so that the scent will disperse more widely.

For social communication within the pack, facial expressions, body language, and tail position all have meanings. For instance, the alpha male and female hold their tails straight out, while subordinate wolves have drooping tails. Dominant wolves use threat displays such as growling, baring their teeth, and raising their hackles. A submissive wolf, on the other hand, may whine, tuck its tail between its legs, or roll onto its back. This sort of communication helps to keep a social order in the pack, and probably prevents a great deal of outright fighting.

Wolves are extremely territorial and will fight fiercely to defend their territory against other wolves. Indeed, it isn't uncommon for wolves to be killed in these territorial battles. This is especially true in the case of lone interlopers, who may be pursued and killed by the whole pack. In some areas these territorial defenses even constitute a major source of wolf mortality.

Much of a pack's time is spent in hunting, for wolves are almost totally carnivorous. Although they catch smaller creatures, such as ptarmigan, hares, and mice, wolves rely principally on big game for sustenance. Large animals such as deer, elk, caribou, and musk oxen are essential to the wolf's survival: the pack can't catch and kill enough small game to feed itself for any length of time, so the large quantities of meat provided by big mammals are a necessity.

Farley Mowat's book *Never Cry Wolf,* and the movie and video versions of it, have duped untold numbers of people into believing that wolves subsist mainly on mice during the summer. I was, I confess, among the "dupees." I first read the book when I was in college and was greatly impressed by it. Mowat is a most entertaining writer, and I accepted the book as fact. Only later did I learn that the book, which defends—indeed, glorifies—the wolf is fictional.

L. David Mech, probably our leading authority on wolves in North America and a scientist who has done an enormous amount to educate the public about this widely misunderstood predator, put it rather succinctly. After noting that no scientist, Mowat notwithstanding, has ever found a population of wolves that regularly subsists on small prey, he states, "However, Mowat is not a scientist, and his book, although presented as truth, is fiction."

A wolf attack on big game is often not a pretty spectacle, despite the illusion that many people hold of the wolf as a swift, silent killer. Although small prey and some large animals—especially their young—are brought down and killed rapidly, the death of many others is a time-consuming and terrifying affair for the prey. It's also difficult and dangerous for the wolves themselves, and biologists estimate that less than 10 percent of wolf attacks on large mammals are successful.

Wolves test different animals until one of the alpha wolves senses weakness on the part of a particular specimen. If the prey, particularly a moose, stands its ground, the wolves will leave it. Large animals that are strong, healthy, and stationary present a serious danger to wolves. A slashing hoof can break ribs or crush a skull, and a sharp antler or horn can gore or impale. The attack isn't worth the risk.

Prey that flees is a different matter. The wolf pack pursues, trying to gauge the strength of the animal. If it fails to gain on the animal within a few minutes, the pack usually abandons the chase. If the wolves are able to close on their prey, however, they begin nipping at heels and flanks to avoid fatal blows from front hooves, antlers, or horns. Eventually the animal tires, and more and more bites are inflicted. Finally one of the wolves, usually the alpha male, runs forward and seizes the prey by the nose. Aided by this distraction, the other wolves rush in, attack flanks and throat, and finish the now-helpless animal.

What follows is no sedate, mannerly Sunday-school picnic. As might be expected, the alpha pair usually get the internal organs, which are the richest, choicest morsels. Otherwise, however, pack members growl, jostle, snarl, and otherwise compete for this rich lode of meat. Finally, even the subordinate wolves have their chance, until all have eaten their fill. At such times of plenty, wolves gorge themselves, putting away ten to twenty pounds of meat at a sitting before collapsing for a well-earned rest. Wolves will also cache excess food from a large kill by burying some of the meat, laying food away against a time of poor hunting.

It's true that wolves prey mainly on the weak, sick, and old, since they are the easiest to catch and kill. Over long periods, this behavior benefits the prey species via the immutable law of survival of the fittest. Wolves, however, like most predators, are very opportunistic and kill strong, healthy animals whenever conditions are right. They also kill many young animals that, regardless of their inherent genetic quality, simply aren't yet old enough or fast enough to elude them. Thus the notion that wolves kill only the old, weak, sick, and unfit, often advanced by those who either don't understand wolves or are trying to glorify them unduly, is a substantial distortion of the truth.

Breeding season takes place in late winter or early spring, depending on the latitude. As might be expected, breeding takes place earlier in more southerly latitudes, and later in the harsher climate of more northerly reaches. Most breeding is done by the alpha male and female. This pair often mates for life, though it's by no means a universal practice. Alpha males are known to breed the alpha female plus a subordinate female, and occasionally to switch mates. Further, if something happens to its mate, the surviving alpha wolf will find another mate. Sentimental twaddle about wolves grieving for a lost mate and thereafter remaining celibate can be dispensed with.

Breeding of a second, subordinate female usually occurs when prey is exceptionally abundant. Even with plenty of prey, however, the pups borne by the subordinate female often don't survive.

Although wolves usually sleep in the open, the female has her litter in a den. There are usually five or six pups, which weigh a pound apiece and are both blind and deaf at first. The warmth of the mother is vital for the pups during their first three weeks, so she's mostly confined to the den during that period, and is fed by the alpha male and other pack members, who consume food elsewhere and regurgitate it for her.

At about two weeks, the pups' eyes open, and they can hear when they're about three weeks old. Then, by the time they're a month old, they begin to leave the den. They begin to eat solid food, and pack members now begin to bring food for the fast-growing pups.

When the pups are about two months old, they're taken to a rendezvous site. This is a location where the pups can be left safely while the pack members go forth to hunt, then rendezvous at the spot. A rendezvous site features some sort of protection for the pups, such as a crevice in the rocks or extremely dense vegetation. There the pups remain, sometimes exploring for short distances around the site, while the rest of the pack hunts, returning periodically to the pups.

Then, by fall, the pups are big enough to travel and hunt with the pack, and they're nearly full-grown by the time they're a year old. Although they're capable of reproducing by age two, full sexual maturity isn't achieved until age four or five.

Nowhere do the conflicting views of wolves display themselves more starkly than over the issue of wolf reintroduction. Under the Endangered Species Act, the U.S. Fish and Wildlife Service (USFWS) held numerous public meetings and hearings in the West, and received testimony in various forms from thousands of citizens. Although predictably polarized for and against wolf restoration, comments heavily favored the former. As a result, the USFWS trapped wolves in Canada and released them at two locations, Yellowstone National Park and the Frank Church–River of No Return Wilderness Area in central Idaho.

Biologically, these transplants have been a howling success (pun intended). There are now more than 150 wolves in a number of packs in the central Idaho area, while the Yellowstone wolf population, counting this year's crop of pups, stands at about 350. Alpha wolves have mated, and packs have formed in normal fashion, fought over territory, killed each other in territorial battles, and otherwise behaved in similar fashion to wolves in areas where they've existed for tens of thousands of years.

Politically, the road has been far rockier. To minimize the fears of ranchers that wolves spreading out from these release sites would decimate their livestock, the USFWS declared these wolf reintroductions "experimental populations." Under the terms of the Endangered Species Act, this designation allows the flexibility to remove or kill wolves that are preying on livestock and pets.

Two diametrically opposed groups then brought suit to halt and reverse the wolf reintroductions. On the one hand was the American Farm Bureau, unalterably opposed to wolf reintroductions; on the other was a consortium of groups so pro-wolf that they felt the experimental population designation didn't give sufficient protection to wolves that might naturally disperse into these areas from Canada.

In December 1997 a federal district court judge ruled that the experimental-population designation was illegal because it gave insufficient protection to wolves naturally dispersing into Yellowstone and central Idaho. The judge ordered all reintroduced wolves *and their offspring* to be removed, then stayed his order so that appeals could be filed.

After numerous delays, as of this writing the appeals have yet to be heard. Meanwhile, nearly two years have passed and many more pups have added greatly to the wolf population in these two areas. The judge's decision was based on what many observers regard as a bizarre interpretation of the Endangered Species Act. Under his interpretation, a dispersing wolf here and there constitutes a "population," a definition that few believe will be sustained on appeal. Further, the only way to get rid of the several hundred wolves now present would be to kill them, since trying to trap and relocate them would be an extraordinarily difficult task at this point. The likelihood that the American public at this point would stand for the slaughter of all those wolves is about as great as the chances of that famous proverbial snowball!

Depredation on livestock has been relatively light, and lower than projected by the USFWS. Most such depredation remains inordinately controversial, however, because of the antipathy which some hold toward wolves. To put the matter in perspective, losses from wolves in the greater Yellowstone area have been less than one-thousandth of annual losses of cattle and less than one-hundredth of sheep losses. Indeed, a few ranchers have even testified that their losses have diminished with the advent of the wolves. Why? Because wolves largely displace coyotes, and wolves are easier than coyotes to keep away from the sheep.

Several steps have also been taken to compensate ranchers for livestock depredation by wolves. First, problem wolves are either trapped and transported elsewhere or killed. Second, ranchers grazing livestock on public lands have been granted reduced grazing fees to compensate them for such losses. And, third, Defenders of Wildlife, a private organization, reimburses ranchers for losses proven to have been caused by wolves.

It should be noted that trapping and removing problem wolves hasn't proven very successful. Once wolves have learned how easy it is to kill livestock, they often return or seek other areas where they can prey on cattle and sheep, and these wolves ultimately have to be killed. Moreover, many translocated wolves often die. In either case, most of the problem wolves end up dying soon after their removal. It appears, therefore, that the best solution to problem wolves is simply to kill them in the first place.

Although livestock losses to wolves have been fewer than expected, an occasional rancher has suffered financial loss despite safeguards and compensation. A few have complained that they couldn't locate livestock that they suspected was killed by wolves, or couldn't find it in time to prove that wolves were the culprit, so that they could be compensated. This problem is being addressed by putting a radio tag on each calf; when a calf dies, it can be found quickly enough to see whether or not compensation is warranted.

Despite some complaints, the system of killing problem wolves and compensating farmers for losses has worked quite well in Minnesota for a number of years. Clearly, adoption of this system in the West isn't problem-free, and undoubtedly needs tweaking a bit here and there to fit local conditions, but the difficulties should be manageable in a way that fairly compensates ranchers for losses and yet allows wild populations of wolves to exist.

Now the Northeast seems to be the next likely source of wolf controversy. The U.S. Fish and Wildlife Service is preparing a wolf recovery plan for the region—a plan that could ultimately recommend the release of wolves in remote areas, most likely in northern Maine and in the Adirondack Mountains of New York. Predictably, heated opposition is already lining up, even though the lengthy planning process has hardly begun.

Reaction from New Hampshire was swift, and the legislature passed a law banning the release of wolves within the state. At least part of this reaction was based on fear and sheer ignorance; one legislator even raised the ludicrous specter of packs of wolves attacking tourists and hikers at night! Some farmers' and sportsmen's groups are also opposing any wolf release in their respective states.

On the other hand, thirty-one environmental and conservation groups have formed the Eastern Timber Wolf Recovery Network. These include the National Wildlife Federation, which is the nation's largest private conservation organization, and in aggregate they form a powerful pro-wolf lobby. They have their work cut out for them, however, as the U.S. Fish and Wildlife Service has said that it won't release wolves on private lands over the objections of the owners. With relatively few large blocks of public land in the Northeast, this policy would greatly complicate any wolf release.

The whole controversy could become academic, however, if wild wolves from Canada infiltrate the northern reaches of the Northeast. Two animals—one definitely a wolf and the other quite likely a wolf—have been killed in northern Maine. Moreover, a Maine hunter and woodsman has seen what he firmly believes is a wolf, and has extensively tracked two or three other equally large animals that he feels confident are wolves.

Some biologists believe that the pattern of infiltration from Canada observed in the West is repeating itself in the Northeast. First there were numerous reports of wolf sightings, they say, and then a pack eventually formed. Others feel that the heavily settled St. Lawrence River Valley, the four-lane highway running through it, and the river itself constitute a nearly insurmountable barrier against any significant southward movement of wolves. An occasional wolf, they assert, won't be enough to establish a pack in the Northeast.

Only time will tell how conflicting views among the public and contrasting opinions among biologists will ultimately be resolved. The only thing that seems certain is that the budding uproar over wolf restoration in the Northeast is likely to grow a great deal more heated!

No chapter about wolves would be complete without a warning about wolf-dog hybrids. These crosses between wolves and dogs have become increasingly popular—indeed, something of a fad—within the past few years. Although there are exceptions to any rule, owners of wolf-dog hybrids generally fall into two broad categories. The first is the macho type who wants a pet with a macho reputation to match, while the second is the person enamored with the notion of wildness who wants somehow to own a piece of that wildness.

Let me be very blunt about it: wolf-dog hybrids are DANGEROUS! Many owners of these crosses become highly incensed at the idea that their lovely hybrid could cause problems, let alone harm anyone. However, the facts tell a very different story.

The danger is greatest for small children, roughly a dozen of whom have been killed by wolf-dog hybrids in the past few years. Anyone with any kind of disability, even a temporary one, may also be at risk. Wolves by nature constantly try to assert their dominance, and they're quick to sense weakness in humans; this may lead them to attack in an attempt to become dominant, just as they would with other wolves.

Small children, on the other hand, may for some reason or other appear as prey to a hybrid and awaken its killing instinct. No one knows for certain what triggers these attacks on either adults or children by wolf-dog hybrids, but the frightening thing is that they're usually lightning-fast and totally unexpected. Very often, attacks are by hybrids that have never before acted vicious in any way.

As a result of numerous attacks, a number of states have passed strict requirements for ownership of wolf-dog hybrids. Typically, the hybrid must be leashed and kept behind heavy-duty fencing that extends a specified distance below the surface of the ground. This latter precaution is to keep the hybrid from digging its way out.

It's not only the very young or the infirm who have problems with wolf-dog hybrids, either. Wolf expert Elizabeth Duman put it this way: "Very simply stated, any animal that is very much wolf is going to exhibit enough wolf behavioral traits to warrant special handling." This dictum is borne out nearly every day, as thousands of people who loved the idea of a wolfish pet find that they can't handle its wolfish behavior.

Some are bitten, though usually not seriously harmed, while others can't control the animals that are ruining their homes. As a result of these and other problems, humane societies, organizations dealing specifically with wolves, and similar groups are inundated with problem hybrids, most of which have to be euthanized. Even though some wolf-dog hybrids are, and remain, sweet-tempered and biddable, the majority of them cause problems. Don't acquire one!

Despite the controversy surrounding wolf reintroduction, it clearly is feasible and, many experts would say, desirable in a few places in the United States. Two conditions must be met, however. First, there must be a large, uninhabited area: there are simply too many problems connected with wolves in proximity to humans, their livestock, and their pets to attempt reintroduction in settled areas. Second, wolves that begin preying on domestic animals will often have to be killed. Trapping these individuals and moving them to new locations fails most of the time. Although many wolf lovers bitterly criticize any killing of wolves (the wolf-as-sainted-predator syndrome), those who value this great predator for what it is and are deeply committed to its survival recognize the necessity of this step. As L. David Mech, who has done so much to further wolf survival, summarizes it, "It is unfortunate that some wolves will have to be killed. However, this should be regarded as the necessary price for allowing wolves to live elsewhere."

With attitudes toward wolves rapidly shifting toward the Good Guy image, even if they sometimes become sloppily and irrationally sentimental, it seems likely that we'll see more wolves, at least in certain parts of the United States. The real key to the future of this fascinating great predator is habitat preservation. Take away the wild areas, and wolves will vanish. Preserve large blocks of wilderness, and wolves will thrive.

19

Cousins with a Difference: The Bobcat and the Lynx

MYTHS

- The lynx is larger and fiercer than the bobcat.
- Bobcats scream horribly.

IF THESE TWO CATS WERE HUMAN, THEY WOULD SURELY BE REGARDED NOT MERELY AS COUSINS BUT AS *FIRST* COUSINS. So closely related are they that, while most other members of the world's cat family are lumped into the plebeian genus *Felis,* the bobcat and Canada lynx—plus the nearly identical Eurasian lynx and the Spanish lynx—share their own exclusive genus of *Lynx.* However, just as human first cousins can display great variation, these two short-tailed cats differ widely in build, temperament, habitat requirements, and prey.

Depending on the authority one consults, there are either three or four species worldwide in the genus *Lynx,* and thereby hangs a tale. Scientific names are formulated by scientists known as taxonomists. Taxonomists are specialists in the science of classification; they decide, for example, which family and genus a creature belongs to, and whether or not it should be listed as a distinct species. As in all forms of scientific endeavor, there are differences of opinion among taxonomists, so they may alter a scientific name one year and change it back again another.

This sort of Ping-Pong with the scientific nomenclature of cats has been in full swing in recent years. For a long time the North American lynx was *Lynx canadensis,* while its bobcat cousin was *Lynx rufus.* Then, several years ago, these two cats lost their distinctive genus and were "downgraded" into the far broader genus *Felis,* becoming, respectively, *Felis canadensis* and *Felis rufus.*

Bobcat *(left)*; lynx

But, as the song goes, "The cat came back, it wouldn't stay away," and more recently the bobcat and lynx were restored to their former lofty status as the proud possessors of their own highly exclusive genus. That's not quite the end of the tale, though.

As one biologist put it, taxonomists can be roughly divided into "lumpers" and "splitters." In this instance, the lumpers believe that the North American lynx and the Eurasian lynx are a single, circumpolar species, *Lynx lynx.* The splitters, on the other hand, prefer to call our North American lynx *Lynx canadensis* and the Eurasian brand *Lynx lynx.* In either case, the bobcat remains *Lynx rufus,* and the only other member of the lynx genus, the Spanish or Iberian lynx, continues as an apparently noncontroversial *Lynx pardinus.* While this debate over names is of absorbing interest to many humans, it seems doubtful that the affected members of the cat family are much concerned about it!

While we ponder the scientific names of these two wild cats, a brief digression seems warranted regarding the scientific name of the house cat—that household pet so familiar to us all. Ever since I can remember, the house cat bore the appellation *Felis domestica*—the domestic cat—but many scientists felt that this name was rather ill-suited to the house cat's notably independent nature.

As one authority expressed it, "The house cat isn't domestic—it's *commensal.*" The term comes from the Latin *mensa,* table, and in this instance means that although house cats deign to share our food, they steadfastly refuse to be owned or told what to do. In short, they aren't truly domesticated. It's certainly no accident that the saying "it's like trying to herd cats" has become a metaphor for a virtually impossible task. In somewhat belated recognition of this rather autocratic bent, taxonomists have now awarded the house cat the truly wonderful title of *Felis cattus.* What could be more fitting?

Members of the genus *Lynx,* however, are most certainly neither domestic nor commensal. Thoroughly wild, these elusive cats are seldom seen or heard. Nevertheless, the lynx and bobcat are important components of our North American wildlife, worthy of deep appreciation for their unique qualities.

The widespread impression that the lynx is larger and fiercer than the bobcat (also known as the bay lynx) is erroneous. The lynx certainly *appears* much the larger of the two; its extremely long legs, huge feet, very thick coat, large jowl tufts, and long ear tufts all lend it an aura of size. When stripped down, though, the lynx is the mammalian equivalent of the owl—surprisingly little weight in comparison to apparent size.

Most lynx weigh in the range of fifteen to thirty pounds, with the low end of that range not much bigger than a large house cat. And, despite all appearances to the contrary, the average bobcat actually outweighs the average lynx. Most bobcats weigh between fifteen and thirty-five pounds, although a very large male will sometimes top forty pounds; as with the lynx, however, many bobcats aren't much heavier than a big household tabby.

In temperament, too, appearance belies reality. The lynx looks much larger and fiercer than the bobcat, but whenever the two cats encounter each other in the wild, it's nearly always the less aggressive lynx that gives way.

All of this makes considerable sense when habitat and prey requirements of the two species are considered. The lynx is very much a creature of the boreal forests, where winters are long, the snow soft and deep, and the cold fearfully penetrating. From the Arctic portions of Alaska and the Northwest Territories to all but the northernmost extremity of Labrador, it roams the vast wilderness and endures the harsh winters of the far north. Only in northern bits of the Great Lakes states, parts of the Rocky Mountains, some of the Cascades, and possibly the farther reaches of Maine does this northern denizen reside in the lower forty-eight states.

The lynx's most prominent features come into sharp focus as wonderful adaptations for survival in this unforgiving northland climate. Consider its fur coat, for example—the pelage that makes its owner look so much larger than it really is. Because of this coat, the brutal Arctic cold presents little problem for the lynx, provided always that it can obtain enough food to fuel its body. It would be hard to find a more magnificent pelt than the long, incredibly soft and dense fur of the lynx. Insulated by that wonderful covering, the lynx can easily withstand the bitter cold so prevalent throughout its range.

Then there are those huge feet—great furry pads about as large as those of a cougar. These act as snowshoes to prevent the lynx from becoming mired in the deep snow of frigid northern winters. Despite these "snowshoes," however, the lynx often sinks a considerable distance into the dry, powdery snow. Then its long legs come to the rescue.

One of the reasons why the lynx weighs so little in relation to its apparent size is the length of its legs, yet another adaptation for life in the far north. These seem ridiculously long—for a cat, at least—yet they're invaluable for survival: those long legs prevent its body from dragging deeply in the snow.

All of these adaptations come into play when the lynx pursues its principal food source, the snowshoe hare. Although the lynx will prey on other creatures when hares are scarce, the cat's ultimate fate is inextricably bound to that of the long-eared, big-footed hare. Indeed, so narrow are the lynx's food preferences that it dines on hare to the virtual exclusion of everything else as long as hares are abundant. Only out of sheer necessity does the big-footed cat turn to alternative prey.

The aptly named snowshoe hare is notably cyclical, especially the farther north one travels. Hares go through a boom-and-bust cycle roughly every ten years, gradually building up year by year to well-nigh incredible populations in which there seems to be a hare behind every bush, shrub, and tussock. Then, almost overnight, hare numbers plummet so drastically that there's hardly a hare to be found where previously there were hundreds.

The lynx population, with a lag of a year or two, mirrors the population curve of its favored prey. As hare numbers soar, most female lynx produce an annual litter that averages two kittens, and the kittens have a high survival rate. After the hare population crashes, the number of kittens declines dramatically, and the few kittens born rarely survive.

Nor are adults immune to this sudden shift from superabundance to scarcity, for many lynx starve in the aftermath of the hare's virtual, if temporary, disappearance. One facet of this phenomenon remains a puzzle to biologists, however: while numerous lynx starve during the great dearth of hares, some individuals appear to remain well fed. How they manage this feat while so many of their fellows starve is a mystery. Perhaps the survivors are more adept at foraging on a variety of prey, such as ptarmigan, squirrels, voles, mice—even carrion. No one knows for sure, although further research may provide answers to this riddle.

Lynx are by no means suited for long-distance pursuit in the sense that, for example, wolves are, but they'll pursue hares at high speed for moderate distances, and here their long legs and huge feet, which parallel the adaptations of their prey, again prove invaluable. Even in soft, deep snow, the lynx can often run down a hare within a short distance. The contest is hardly one-sided, however, for the lynx frequently fails in its pursuit of the speedy hare.

Lynx, however, like most other cats, are much more stalkers and pouncers than pursuit animals. Their instinct is to conserve energy; a careful stalk and a swift pounce, or an all-out sprint for a relatively short distance, expend far less precious energy than a lengthy chase. Using an even more energy-efficient technique, lynx often simply lie down along hare runs and wait for their prey to come to them.

Lynx mate during January and February—the dead of winter in the northern climes where they reside. Their gestation period is about two months, so the kittens are born, blind and hairless, in March or April. The female gives birth in whatever sort of den the terrain offers. This can be a hollow log, a hole in a tumble of rocks, a space beneath the roots of a large stump, or some similar place that offers shelter from the elements.

Like cats the world over, the male lynx has no interest in family matters once he has mated, and it's left to the female to raise the kittens in single-parent fashion. Though the kittens are helpless at birth, their eyes open after ten days, and they're soon rolling and tumbling much like domestic kittens.

After about twelve weeks the kittens are large and strong enough to accompany their mother during her quest for food. As they follow her, they learn by example how to stalk their prey, pursue it, or conceal themselves beside a run to await the approach of a hare. Then, once they've received this all-important education, they set out to find their own territories.

North American scientists are only now beginning to study the lynx in anything approaching the detail lavished on many other species. This oversight is partly due to the elusive character of the lynx, its generally inaccessible habitat, and its relatively low population levels. Even the most basic question—how many lynx there are, especially in the lower forty-eight states—hasn't been answered.

That situation is changing, however. John Weaver, a research biologist with the Wildlife Conservation Society, has developed an excellent tool for assessing lynx numbers. After watching his captive lynx rub her cheeks against a post, Weaver devised rubbing pads with nails protruding from them. Treated with his secret formula, which includes catnip, and scattered about in lynx country, these pads collect hairs from the big-footed cats, which seem to enjoy rubbing against them. For example, in Montana's Kootenai National Forest, twenty-eight rubbing pads yielded lynx hairs. The trick, of course, is in determining how many different lynx are represented by these samples. Fortunately, DNA analysis has come to the rescue.

Utilizing cats in its research, the National Cancer Institute has done considerable work on leukemia. In 1995, Stephen O'Brien, performing studies at the Institute, refined a basic DNA analysis so that it became useful in testing lynx hairs. This test is currently being used by the Wildlife Conservation Society to analyze lynx hair samples from the Kootenai Forest and several other locations in Canada and the United States. Preliminary results indicate that several different lynx deposited the twenty-eight hair samples from the Kootenai, but additional analysis is required to determine the minimum number of individuals represented by the samples.

It's estimated that there are several hundred lynx in the lower forty-eight states, but these are often in isolated populations, beleaguered by destruction or fragmentation of their habitat. With pressures of this sort increasing, concerned scientists and conservation organizations have pressed the U.S. Fish and Wildlife Service to list the lynx as threatened or endangered under the Endangered Species Act. The Service has so far declined, citing its limited resources and the need for further lynx studies.

However, biologists with the Colorado Division of Wildlife have released thirty-seven lynx imported from Canada and Alaska. If enough of these survive, the biologists hope to release fifty more of the cats next year. If successful, this would be a major step toward returning the lynx to some of its former range.

Farther north, the lynx is in much better shape, and is thought to be holding its own. Although their numbers fluctuate widely according to hare cycles, there are believed to be at least tens of thousands of lynx in Canada and Alaska.

The life of a bobcat is very different from that of the lynx, close cousins or not. Life in the wild is never easy, but bobcats mostly reside where the climate is far less forbidding than that which confronts the lynx. Indeed, the ranges of the two species scarcely overlap: the bobcat, which roams down through Mexico, barely spills over into southern Canada, while the lynx inhabits only a tiny chunk of the United States, exclusive of Alaska. This rather neat geographic division has its advantages for the lynx: throughout most of its range, it has no competition from the more aggressive bobcat.

With bitter cold and deep snows less of a problem, the bobcat has evolved in a very different manner from its boreal cousin. It features smaller feet, much shorter legs, a more compact frame, and an altogether more typically catlike appearance than the lynx. Its coat, while ample for winter protection in the northern parts of its range, falls far short of the lynx's incredibly thick, luxurious pelt. Even the bobcat's markings are more catlike; its pronounced facial patterns closely resemble those of the tiger tabby model of the house cat, as well as those of a number of the world's small-to-medium wild cats.

If the lynx is a specialist, bound by evolution to the snowshoe hare, the bobcat is a supreme generalist. Rabbits, hares, mice, voles, squirrels, grouse, small birds, birds' eggs, snakes, frogs, crustaceans, and dead—though not rotting—animals are all grist for the bobcat's mill. Although some of its prey are cyclical, the bobcat has a wide array of alternative sources to keep it fed.

Only in severe winters in the northern extension of its range is food likely to be a serious problem for the bobcat, whereupon it often turns to the carcasses of dead deer to carry it through. Bobcats are also known to kill deer during particularly harsh winters. Small deer, less than a year old, are usually its victims, but sometimes older deer are also killed, usually by very large male bobcats. Despite these occasional depredations, the bobcat doesn't rate as an important predator of deer.

Even more than its lynx cousin, the bobcat is a stalker and pouncer. A bobcat may sprint after prey for short distances if the effort seems warranted, but that's not its usual hunting technique. Bobcats are noted for their slow, cautious stalking, and one will sometimes remain for considerable periods beside a burrow, waiting for its unsuspecting prey to emerge.

A friend of mine, Paul Kress, witnessed an amazing demonstration of a bobcat's slow, patient stalking technique, as well as its uncanny ability to pinpoint the location of a fairly distant sound. Turkey hunting in Massachusetts, Paul was seated on a log, dressed in camouflage clothing, using a turkey call from time to time.

A movement caught his eye, and he was astonished to see a bobcat creeping ever so cautiously in his direction. Paul froze, keeping absolutely still in order to avoid alerting the bobcat. On came the bobcat, step by slow step, until it actually started up the log on which Paul was sitting.

At that point the cat seemed to sense that something wasn't exactly as it should be, although it seemed unsure why. It halted, then very slowly backed away before gradually melting into the forest. Even though Paul had stopped calling as soon as he saw the bobcat, the cat seemed to know the precise spot from which the supposed turkey calls had emanated!

Because they feed on such a wide array of creatures, bobcats are far less demanding than lynx in their habitat requirements. Everything from large, wild tracts of northern forest to farm country to scorching desert serves as home for this versatile cat. No doubt this versatility accounts, at least in part, for the bobcat's relative abundance. Estimates of North America's bobcat population vary widely, but even the lowest are nearly three-quarters of a million, while the high estimates are double that.

The preferred denning sites for bobcats are crevices in rocks or ledges, wherever these are available, but hollow logs, pockets beneath blown-down trees, and similar places also serve very well. In usual cat fashion, bobcat kittens are born with their eyes closed. The average litter size is two, but sometimes there are as many as four. Like the lynx and other cats, the kittens open their eyes after about ten days, and begin to forage with their mother after a few weeks. The kittens sometimes leave their mother in the fall, but may also remain with her through the winter. In the southern portions of its range, the bobcat may have two litters a year.

Both bobcat and lynx are wary, elusive, and seldom seen. This trait is much more evident in the bobcat, however, because it frequents far more settled country than the lynx usually inhabits. The bobcat's mainly nocturnal ways add to the difficulty of seeing one, and many people who have spent much time in the woods for years have never laid eyes on a single bobcat. Thus I count myself as extraordinarily fortunate to have had one memorable encounter with this elusive feline.

It was Thanksgiving Day, 1971. I was deer hunting, perched in a tree overlooking an old field that had grown up to tall grass and, here and there, scrubby little bushes. It was over two hundred yards to the woods at the lower edge of the field, and one of my hunting partners was ensconced another hundred yards or so down in the trees, over a steep bank.

The ground was still bare, but the lowering sky was a dark, forbidding gray as the first few flakes of snow began to fall. As I looked toward the bottom of the field, I caught a brief sight of a smallish, stocky animal. At that distance it resembled a squat little dog, and the thought crossed my mind that someone's Scottie might be running loose. Then the snow began to descend in earnest, and I completely forgot about the animal. Faster and faster came the whirling flakes, and in an hour or so there were a number of inches of snow on the ground, while every little shrub and tussock was draped in a thick mantle of white.

Only twenty or thirty yards below me was a path where my hunting partner had trampled the long grass on his daily trips to his chosen hunting spot. Without warning or any prior hint of movement, a bobcat materialized in the path and stood there, almost facing me. I was stunned by its sudden appearance, but had presence of mind enough to raise my rifle and view the cat through the scope.

Several thoughts ran through my mind as the scope revealed the cat in wondrous detail. The first was what a beautiful animal it was. The second was that its bold facial markings were almost identical to those of our two gray tiger tabbies. The third was that it was one very cross-looking cat, seemingly at odds with the whole world! I marveled at the sight—and then, as if without motion, the cat dematerialized as mysteriously as it had appeared.

Some time later I made out the form of my hunting partner, trudging uphill through the swiftly deepening snow. I hurried down to meet him, and excitedly asked, "Guess what I saw, Art?"

"A bobcat," he replied.

"How did you know?"

"Well," he informed me, "just as it was starting to snow, I heard a noise in the leaves up the bank from me. I turned, and there were a mother bobcat and three kittens that were about three-quarters grown. They saw me and scattered in all directions."

It was the mother that I had seen, of course; the kittens had evidently stayed in the woods below me and hadn't come into the field with their mother. Probably the reason why the cat looked so cross was that she and the kittens had been temporarily separated, though they no doubt found each other soon after. Incidentally, we got twenty-two inches of snow that day, making it my most memorable Thanksgiving in more ways than one!

Our younger son also saw a bobcat, but under very different circumstances. As he was driving through a wooded area, a bobcat raced out of the snow-covered evergreens into the path of the car ahead of him. Dave immediately stopped to check on the cat, which was unfortunately dead. It was unmarked, however, as it had apparently been struck in the head, but not run over. It was a beautiful specimen, so Dave took it to the Vermont Institute of Natural Science, where, the last I knew, it was preserved in their freezer.

Then there was the very strange experience of an acquaintance, who lives on the fringe of Montpelier, Vermont's capital city. He looked up from his yard work to see a doe walking down the street, followed at a discreet distance by a fairly small bobcat! Why the bobcat was acting in that fashion is something of a mystery. Certainly it wasn't about to attack a full-grown deer in warm weather. Quite possibly it was a kitten, almost grown and just setting out on its own, either testing its stalking skills or simply following the much larger animal out of sheer curiosity.

Both bobcat and lynx can be quite vocal at times, although they don't make a habit of it. The bobcat's voice is often described as sounding very much like that of an oversized house cat, with loud yowling, growling, hissing, and spitting. Certainly my only experience with a bobcat's voice reinforces that description.

One frosty autumn morning while I was in high school, I arose at first light to do a little hunting before school. I was crossing the field behind our house, when from the nearby woods came loud noises that sounded for all the world like a solo version of a fight between house cats, only louder. Although I suspected the source, I wasn't sure until a neighbor who was an old woodsman, hunter, and trapper emerged from the woods.

I hurried over to him and asked, "Did you hear that noise? What was it?"

"Bobcat," was his succinct reply.

Over the years, however, I've encountered a number of people who have recounted how frightened they were, almost always at night, by "a bobcat's scream," which they generally described as "bloodcurdling" and "sounding like a woman who's being murdered." This simply doesn't seem to fit with what's known about the bobcat's vocal repertoire: perhaps it's possible to interpret the sort of *raowraowraowraow* yowls, such as fighting tomcats unleash on each other, as bloodcurdling screams, but this seems a bit far-fetched. The likeliest explanation is that the high-pitched scream uttered by none other than our old friend the barred owl (see page 110) is frequently attributed to the bobcat.

Bobcat populations generally seem to be holding their own, and may even be increasing. Indeed, this versatile feline is full of surprises: despite its generally reclusive nature, bobcats of late have been adapting to life in the outskirts and suburbs of a number of cities. Although they still tend to avoid being seen during daylight hours, or walking where they're readily visible, bobcats have apparently adjusted to the proximity of large numbers of people. Such adaptability, of course, bodes well for the future of this handsome feline.

While there's justifiable concern about the future of the lynx in the southern extremities of its range, the big-footed cat appears to be doing well in Canada and Alaska. Although its cyclical nature makes population trends difficult to determine, it seems likely that the lynx, together with its bobcat cousin, will continue to intrigue and delight all those who find the wondrous abilities of cats so fascinating.

20

The Magnificent One: The Cougar

MYTHS

- The cougar is an endangered species.
- Cougars are no threat to people.

THE COUGAR *(PUMA CONCOLOR)* MIGHT WELL BE DUBBED THE CAT OF MANY NAMES. Mountain lion, puma, catamount, panther, and painter are among the most common, but numerous other English, Spanish, and Native American titles have been applied to this big cat. The sources of these names are as varied and curious as the names themselves. *Mountain lion* is self-explanatory, but *puma* comes through Spanish from a Peruvian word. *Panther* entered the English language from the Old French *pantere,* which in turn goes back through Latin *panthera* to the Greek *panther. Painter* is simply a colloquial pronunciation of panther, and *catamount* is a form of *cat o'mountain.* And *cougar,* my personal preference, is French, taken from the language of the Tupi Indians of South America.

"What's in a name?" Shakespeare wrote. "That which we call a rose by any other name would smell as sweet," and this gorgeous great predator is just as splendid whether we call it cougar, puma, mountain lion, panther, or some other name. This is by far the largest cat commonly found in North America above the U.S.–Mexican border; although the jaguar is slightly heavier, it rarely reaches the U.S. side.

The cougar's scientific name, *Puma concolor,* means a puma of the same—or one—color (until recently, the name was *Felis concolor*—cat of one color). At any rate, the species name is an apt description of this big feline's basic appearance. Most of the cougar's coat is tawny, reminiscent of the African

Cougar

lion's color, but the underparts are much lighter, and the tip of the tail and backs of the ears are dark brown to nearly black.

Much of the cougar's arresting beauty stems from its opposite ends—its face and tail. Viewed face on, the cougar's tan head culminates in a strikingly handsome pattern: its pink nose is surrounded by a white muzzle, set off by vertical blackish bars at the rear. At the other end, the cougar's tail is perhaps its chief glory. Nearly three feet long on really large specimens, this appendage, when adorned by the cougar's winter fur, is truly majestic. Indeed, in its thick winter pelage, the cougar fully deserves to be called magnificent.

Just how large are cougars? The heaviest recorded specimen, a male, weighed a whopping 275 pounds, but that's an obvious anomaly. Other male cougars, which are substantially heavier than females, have occasionally reached 225 to 230 pounds, but John Beecham, a cougar expert with the Idaho Fish and Game Department, says it's extremely rare to encounter a male over 200 pounds; 130 to 150 pounds is more typical for adult males, while females tend to weigh seventy-five to ninety pounds.

As for length, big males can run seven to eight feet from nose to tip of tail. In one study, adult males averaged just over seven feet in total length, females about six and a quarter feet. The cougar's splendid tail accounts for between 35 and 40 percent of that total, although it often appears longer.

For reasons that aren't entirely clear, there seems to be a pervasive feeling that cougars as a species are—depending on whose opinion one listens to—either rare, threatened, or endangered. Perhaps this is due to the publicity given to the Florida panther *(Puma concolor coryi),* a cougar subspecies that is genuinely endangered; after all, the image of a Florida panther on a poster or an endangered-species stamp looks very much like any other cougar. Also, the fact that cougars are listed as endangered in the eastern United States may lead people to think that the species overall is in deep trouble.

In truth, cougar populations in the western states and parts of western Canada have increased greatly in the past couple of decades and seem to be thriving. But how many cougars are there? Unfortunately, that's a very difficult question to answer. Counting cats, it turns out, is only slightly easier than the proverbial task of herding them!

Like most wild cats, cougars are highly secretive. Further, many of them live in the remote vastness of wilderness areas. This makes any accurate assessment of their numbers a daunting task. Nonetheless, based on the best estimates possible under the circumstances, biologists have come up with a total figure of 25,000 to 30,000 cougars in the United States and Canada. (These figures should be viewed in light of the biologists' caveat that they are *very* rough estimates.)

Cougar numbers in the East (excluding Florida) are another matter. Cougars were long ago eliminated from the East, but have they returned to their old haunts? Many people think so. Although the idea of a thriving eastern cougar population is at best highly speculative, it's nonetheless hotly debated. Reports of cougar sightings are rife in the northeastern states, and several organizations exist exclusively to promote the belief that thousands of cougars roam the forests there.

Biologists point to the almost total lack of any hard evidence that these great cats are present as a breeding population in the Northeast. They believe that most cougar sightings are cases of mistaken identity, and those that are genuine are mostly "pet" cougars released into the wild.

It's an unfortunate fact that cougar kittens can easily be purchased from animal farms in some states. Emotion overcomes some people when they view these undeniably adorable kittens and fail to foresee the day when their call of, "Here kitty, kitty, kitty," is answered by a cat weighing one hundred pounds or more! Ultimately, owners who don't know what else to do with such a large and potentially dangerous cat simply take it to a tract of forest land and release it. The tragedy is that cougars raised by humans almost certainly lack the hunting skills to survive for long in the wild.

True believers scoff at the notion that most alleged sightings are cases of mistaken identity, or that valid sightings are of "pet" cougars released into the wild. Further, they accuse wildlife biologists of conspiring to hide the truth because they don't want an endangered species on their hands.

This view is both unfair and farfetched. Most biologists are fascinated by the big cats and would welcome the opportunity to study them firsthand. Further, it seems highly improbable that large numbers of cougars could inhabit the area without leaving tangible evidence of their presence, such as tracks, droppings, and deer kills. Cougars are certainly secretive and tend to keep out of sight, yet the endangered and seldom seen Florida panther—of which only thirty to fifty remain—leaves ample evidence of its existence.

Among other things, road-killed cougars are by no means unusual where cougars are known to exist. There are at least sixteen documented road kills of the endangered Florida panther in the last two decades, yet not a single road-killed cougar has been reported elsewhere in the eastern United States.

Another puzzling phenomenon is that a high percentage of supposed cougar sightings are reported as "black panthers." The true black panther is a melanistic phase of the leopard, but there has never been a single documented case of a melanistic cougar anywhere. However, in the minds of a substantial portion of the public, the term "black panther" is firmly rooted, and so many black animals of various sorts are reported as cougars, alias black panthers.

This brings us to the amusing and highly instructive tale of a "black panther" that recently caused a furor in Vermont. A nurse leaving a nursing home saw what she believed was a black panther. She grabbed her camera and snapped several pictures, which she then took to the Fish and Wildlife Department. Department personnel visited the site and tried to estimate the size of what was clearly a black cat, based on the size of background objects. Their conclusion was that—in addition to being black—the cat was too small to be a cougar, but that it might conceivably weigh forty pounds and be the melanistic phase of some smaller species of cat released into the wild. Photos of the cat appeared in major newspapers, and the whole affair became the talk of much of the state.

Three or four days later, a lady who lived close to the nursing home took Porgy, her black male house cat, to the veterinarian for a routine checkup. The veterinarian, who had seen the newspaper photos, looked at Porgy, then said to the woman, "Don't you think you'd better 'fess up?" She admitted that she ought to, but had been reluctant to do so because of all the publicity. She then contacted the Fish and Wildlife Department and told them that the photos were indeed of her wandering tom, who weighed less than twenty pounds, identifiable from the photos by a small white spot on his throat. Thus did a house cat named Porgy become a media star and "panther for a day"!

Erroneous reports of cougar sightings are by no means confined to the East. Cougar research biologist Paul Beier, who works in southern California, estimates that at least 75 percent, and perhaps as many as 95 percent, of reported cougar sightings are cases of mistaken identity. An astonishing range of animals, from domestic dogs and coyotes to bobcats and deer, are erroneously identified as cougars, and even experienced observers have fallen into this trap.

To date, only two cougar sightings—one in Maine and one in Vermont—have been validated (the one in Vermont by DNA testing of a hair sample). There has been no further evidence of these cougars, and it must be presumed that they were released specimens that have since perished. Beyond that, there is not a single piece of confirmed evidence that cougars exist in the East. If a breeding population were truly present in the East, it's difficult to imagine that not a single cougar would have been struck by a car, or that no one would find deer killed by cougars, identifiable tracks, or other firm evidence of the cat's presence.

The term "magnificent" applies not only to the cougar's appearance, but to its physical prowess as well. In addition to its beautiful appearance, the cougar possesses that innate grace which seems to be nature's gift to the cat family. Coupled with extraordinary power and explosive speed in a short sprint, this makes the big cat a marvelous killing machine that can dispatch prey many times its size. For instance, a 150-pound cougar can quickly kill an elk weighing up to one thousand pounds.

Like most other cats, the cougar isn't a pursuit animal. In typical cat fashion, its usual mode of attack is to stalk its prey until it comes within striking distance. Then, with an amazing blend of speed and power, the cat overtakes its prey, leaps upon it, and delivers a killing bite to the neck, often at the base of the skull. That, at least, is what happens in the course of a successful attack; as with other predators of large animals, the cougar's attacks fail more often than not, though that hard reality in no way diminishes the cougar's exceptional skill as a hunter.

There are many ways in which even the cougar's lightning attacks can go awry. Large prey—deer, elk, bighorn sheep, and the like—are extremely wary. No matter how careful the stalk, the slightest hint of movement will alert the prey and perhaps launch it into full flight. An errant puff of wind may carry the cat's scent to the supersensitive nostrils of its intended dinner, or a tiny noise made by those soft, padded feet can sound the alarm.

Even if the cougar successfully moves into position to commence an attack, many things can still go wrong. The cougar depends upon a swift end to its onslaught, and the prey may be able, for example, to dodge the racing predator and accelerate to the point where the cat gives up the chase. At other times the prey may have sufficient warning to simply be able to outrun its pursuer. Predation, even by a cougar, is far from a sure thing.

Big game—deer, elk, and bighorn sheep—are the mainstays of the cougar's diet; it's also known to kill moose, though usually calves rather than adults. The big cat rounds out its diet with substantial numbers of smaller mammals such as hares, rabbits, grouse, coyotes, foxes, badgers, skunks, opossums, and raccoons. As previously noted, cougars are one of the very few animals able to kill an adult beaver (see chapter 5), which finds its way into the big cat's diet from time to time. Small rodents, including ground squirrels and even mice, are also consumed, and cougars may feed heavily on them when they're available in large numbers. However, cougars, like wolves, don't survive for long unless large prey is common.

When a cougar kills a large animal, it may feed on it for as much as ten days. Often it will drag its kill into woods or brush where it can conceal it, frequently by covering it, to keep its presence secret from scavengers. By the time a cougar finishes with a carcass, there is amazingly little left. Even the large upper leg bones are devoured, and only the jaws, the lower legs below the "knee" and "elbow" joints, and the stomach remain.

This brings up an interesting point. Researcher Kenneth Logan, at the Hornocker Wildlife Institute, has seen a cougar eat a mouse and expel the stomach—something that one would hardly expect from such a large predator eating such small prey. I've noticed that our house cats also leave the stomachs of their prey. Is this a trait shared by all cats? Members of the dog tribe consume the stomachs of their prey, and even clean up those left by cougars. Perhaps this is yet another difference between cats and dogs.

Breeding male mountain lions are highly territorial and keep other males away, while females have overlapping ranges. Thus there is a surplus, nonbreeding cougar population, much as there is with wolves; young male cougars must wait for older males to die or become decrepit with age or injuries before moving into the breeding population.

Cougars only reproduce once every two to three years. Cat-fashion, cougar toms take no interest in raising their offspring; once mating is over, the male and female go their solitary ways. After a gestation of about three months, the females have a litter of kittens, normally two or three but occasionally more. These can be born at almost any time of the year.

As with the young of other cats, the kittens' eyes are closed for the first ten days. The baby cougars are spotted, and remain so until they're six months old. By the age of nine months, the young cougars are capable of hunting on their own, but they usually remain with their mother for up to two years.

One of the most astonishing facets of cougar behavior is their vocal versatility. Few would imagine that this great cat could chirp almost like a bird, yet a mother will use such sounds to communicate with her kittens, for example. They also produce fairly elaborate whistling sounds that go up and down the scale. Other cougar sounds are far less musical. One has been described as a rather raspy "ow" sound, which, along with a sort of throaty coughing noise, is probably about as close as this big cat comes to a true roar. Then there are growls and hisses, much like those of a house cat, though considerably louder.

Even this array of sounds fails to encompass the full range of cougar vocalizations. Females in heat utter what some have described as a piercing scream, which Kenneth Logan prefers to call "caterwauling." He describes this as being much like the cries of a house cat in heat, although considerably louder. Having experienced this caterwauling at close range on a number of occasions, he believes that those who describe it as a high-pitched scream have heard it at a distance, with trees or other obstacles filtering out all but the highest tones. Because cougars are solitary and often live far apart, this loud yowling probably aids the males in locating females in heat.

Finally, cougars have the virtue of being able to purr loudly and contentedly. Logan says he has been in a cage with a purring cougar, and he describes the sound as "thunderous" at that range.

It's long been an article of faith that cougars don't attack people. That was never quite true, but cougar attacks in the past were indeed extremely rare. During the past few years, though, the number of cougar attacks on humans has escalated substantially. Three people have been killed by cougars in Colorado, and a number of others have been mauled or stalked in several states within the past five or six years. Though such attacks still don't represent a major threat to people, they're nonetheless disturbing.

Cougar experts believe the trend toward more cougar attacks has three causes. First, cougar populations have increased greatly in some areas. Second, large numbers of homes, with their attendant human activities, have invaded major blocks of prime cougar habitat. This situation naturally creates many more opportunities for interactions between people and cougars. Third, halting all cougar hunting in California has created cougar populations there that have little or no fear of humans.

None of this should cause people to have an inordinate fear of cougars. The number of cougar attacks on humans is still very small, representing far less of a threat to human life than lightning or bee stings—to say nothing of automobiles. Nevertheless, in view of the increasing number of attacks, it's wise for anyone living or hiking in cougar country to know how to handle a confrontation with one of the big predators. State wildlife agencies in the West usually have valuable information on this subject; the California Department of Fish and Game, for example, has an excellent handout called "Living with California Mountain Lions."

Once widely persecuted as a varmint, and eliminated from the eastern portion of its range, the cougar's future today looks reasonably bright. Although the big cats are hunted as game animals in several states, hunting is closely regulated and in no way threatens cougar populations. The one serious danger facing the cougar is loss or fragmentation of habitat. As humans continue to encroach on the cougar's territory, homes and concomitant development such as roads destroy valuable habitat. Still, with large areas of the West set aside in wilderness, national forests, and other public lands, this magnificent big cat should continue to flourish.

21

Three Minus Goldilocks: Black, Brown, and Polar Bears

MYTHS

- Bears in national parks and similar places are tame and safe to feed or pet.
- Bears eat mostly meat.
- Bears can't run fast uphill, so a person can escape a bear by running up a steep incline.
- Most bears den in caves.
- Hibernating bears sleep so soundly that they don't even awaken when their cubs are born.
- Black bears hoot, almost like an owl.
- Bears often attack people.
- All bears are good tree climbers.
- Alaska brown bears are easily the world's largest bears.
- Polar bears are white.

BEARS ARE FASCINATING CREATURES. Although they don't arouse the hatred and fear engendered by the wolf—after all, the bears in the story of Goldilocks meant well, even if they frightened the little girl—they're often misunderstood and widely underappreciated. They're the largest of all terrestrial predators, immensely powerful and, when they choose to be, very fast for short distances.

Our three North American bears are related quite closely, and share many anatomical features and traits, but they also display surprising diversity in a number of ways, such as size, temperament, diet, and habitat.

Black bear

The Black Bear

The black bear *(Ursus americanus)* and its brown phase, the cinnamon bear, are by far our commonest bears. Current estimates of their population run around 750,000, with approximately 100,000 in Alaska alone, so there's most assuredly no shortage of this species; in fact, black bears are actually increasing in many areas. The black bear is also the smallest of our three native bears: even though a very few extraordinarily big, fat specimens exceed eight hundred pounds, those are no more the norm than seven-foot-two-inch basketball players or 325-pound football linemen are typical of humans. Adult black bears commonly range from 150 to four hundred pounds, with the vast majority three hundred pounds or less.

Normally extremely shy and reclusive, black bears usually flee at the first sign of a human. But these animals can also become habituated to people rather quickly, and therein lies a major problem. Black bears—or, more properly, people fascinated by black bears—are the bane of park rangers and other public lands officials. Despite every warning, ignorant individuals persist in thinking that black bears begging for handouts along roadsides and in parking lots and campgrounds are tame—a sort of hybrid composed of equal parts Yogi Bear, Gentle Ben, and a child's teddy bear. In fact, these bears are extremely dangerous because, while not tame, they've lost their fear of humans.

Every year people are injured, some seriously, when they try to feed bears, despite all the warning signs and lectures. Worse yet, I've heard horrific tales from thoroughly reliable sources about people who have tried to take pictures of their children with a bear—and have even attempted to put their child on a bear's back!

Conflict between black bears and humans in such places as parks and campgrounds is hardly new, although the growing popularity of travel to these spots has certainly exacerbated the situation. In the 1920s, for instance, my mother's cousin, a civil engineer, headed west to work in the mining industry. After he had reached one of the western states, he stopped to ask directions at a ranger station in a national park.

A New Englander unfamiliar with park bears and their ways, he made the mistake of leaving his car door open when he went into the office to talk with the ranger. As he emerged from the ranger station, he saw a hefty black bear in the front seat, industriously rummaging through his lunch on the back seat. Just at that moment, the bear leaned back and pressed against the horn button. The horn beeped loudly, and disaster swiftly ensued, for the alarmed bear catapulted itself straight through the canvas roof of the car and disappeared into the nearby forest!

As more and more cities and suburbs spill out into surrounding bear habitat, an increasing number of people are having conflicts with black bears. Colorado alone has recorded some 450 incidents involving bears and humans during the past year—and most of those have been in urban and suburban settings. Like park bears, these bruins have lost much of their fear of humans and scavenge for garbage, try to break into homes, and otherwise cause serious problems. Although few individuals are actually killed in these encounters, they're sometimes attacked and mauled.

Attacks on humans by truly wild black bears are extremely rare, but a mother bear with cubs can be exceedingly dangerous. Just two or three years ago, a Vermont farmer was attacked by a bear with a cub when he went out in late spring to check his maple sugar woods. When the bear charged him, the farmer hastily climbed a tree. But even though the cub had gone up another tree and was out of danger, the mother nonetheless came up the tree after the farmer.

Black bears are excellent tree climbers, and this one was no exception. Up the tree she came after the farmer, who defended himself by vigorously kicking the female bear—called a sow—in the nose and face, while she bit at his boot. Finally he inflicted enough damage so that the sow backed down, called her cub, and disappeared into the surrounding woods.

The farmer climbed down after what seemed a prudent interval, but the sow, who was only in hiding nearby, immediately rushed him and renewed the attack. Again the farmer climbed the tree and fought her off. When she finally climbed down and disappeared for the second time, the farmer wisely decided to remain where he was. After two or three more hours, some of his family came looking for him, and by that time the bear had departed. While sustained attacks like this are rare, even by a sow with cubs, it does serve to show why a bear with cubs should be avoided at all costs.

Bears are usually regarded as primarily meat eaters, and they're indeed carnivores, but the vast majority of a black bear's diet consists of plant food, from a variety of early spring greens to berries, apples, beechnuts, acorns, and corn in the fall. While all of these are important in their season, the early-spring greens are the most critical for black bears: just emerging from hibernation, the bears require the nutrients that these plants supply. Upland wetlands, where the snow melts early and green plants poke up through the ground while surrounding areas are still snow-covered, are therefore a vital part of good black bear habitat.

As already noted, black bears can zip up and down a tree with great rapidity. In the fall they climb beech trees in search of the highly nutritious beechnuts, which they're exceedingly fond of, and the parallel claw marks that they leave on the smooth, silvery beech bark are clear evidence that bears have been there. Bears also make so-called "bears' nests" in beech trees, although these aren't actually nests. A bear sits in one spot high in the beech tree and reaches out to pull nut-laden branches toward itself. Many branches and twigs are partially broken in the process, for bears are extremely powerful. The result is a tangle of branches that, from the ground, somewhat resembles a huge nest.

Meat sources for black bears are usually somewhat limited. They include fawns in the spring, ants and grubs, and mice and squirrels whenever a bear can dig them out. Frogs and fish are also fair game, and carrion and garbage, when available, are also part of their diet. Bears have a reputation for possessing a great sweet tooth, but they may actually rip open domestic hives and bee trees more for the bees and their larvae than for the honey.

One of the hoarier myths about black bears is that they can't run uphill very well because their hind legs are short, so anyone being chased by a bear should try to escape by racing up a steep slope. Another version holds exactly the opposite to be the case: black bears can't run fast *downhill,* because their hind legs are so short! How these notions ever originated is anyone's guess, but they've been around for a very long time. In any event, both versions are emphatically untrue. Uphill, downhill, sidehill, or on the level, a bear can outrun a person with ease. Although bears aren't long-distance runners, for short distances they can sprint astonishingly fast for what might seem a roly-poly creature. Their top speed is at least thirty miles an hour and probably more—almost as fast as a white-tailed deer can run!

Another great myth about black bears is that they hoot almost like owls. This particular myth seems to be confined to portions of northern New England, especially Vermont, since biologists in other parts of the United States have never heard of it. For years I was told repeatedly that black bears hoot, and I was puzzled and disturbed by my apparent inability to tell which hoots belonged to barred owls and which to black bears. When I inquired of this person or that about how one could differentiate between the hoots of the two, I always received answers that were substantially the same: "They sound a lot alike, but you can hear the difference if you really listen carefully."

I listened carefully, over and over, but no matter how hard I tried, they all sounded like barred owls to me. Then one day I was discussing wildlife myths with Charles Willey, who was the bear biologist for the Vermont Fish and

Wildlife Department at that time. "There's one myth I hope you can correct," Charlie told me, "and that's the one that bears hoot."

"You mean that they *don't*?" I asked him in surprise and not a little relief. He informed me that no reputable biologist or scientist, even those who have worked closely with hundreds of black bears over many years, has ever heard or seen a bear hoot. Black bears make a wide variety of sounds—growls, whines, grunts, rumbles, snorts, woofs, and whuffles among them—but they don't hoot!

Trying to disabuse people of this conviction, however, is a nearly hopeless task. Most who believe this myth embrace it with an almost religious fervor, and utterly refuse to listen to contrary opinions. For example, an acquaintance of mine who's a noted ornithologist told me about emerging from his car at a friend's house, just as a barred owl hooted close by. "That's a bear," said his host excitedly. When the ornithologist demurred and identified the hooter as a barred owl, the host flatly refused to believe him!

Charlie Willey also told me an amusing and instructive anecdote concerning a college student from Vermont who was Charlie's assistant one summer. In the course of their work with bears, Charlie mentioned that bears don't hoot. The young man, who had been raised on this fiction, was somewhat skeptical at first, but finally appeared convinced that Charlie was right. Then he went home for the weekend and returned on Monday to tell Charlie with great excitement that bears really *do* hoot. It seems that he had gone into the woods with an old hunter and woodsman, and at some point they had heard a bear hoot.

"But how do you know it was a bear?" Charlie asked him.

"Because he told me so," responded his assistant. This reply was remarkably similar to that of Josephine in Gilbert and Sullivan's *HMS Pinafore*: "But I know Sir Joseph is a good and great man—for he told me so himself."

Like the brown and polar bears, black bears have delayed implantation of the embryo. Although mating occurs in the summer, the fertile eggs don't implant and begin to develop inside the female until late fall. This is an evolutionary adaptation that makes great sense. In years when food is very scarce before hibernation, cubs might sap the mother's strength so that both mother and cubs would perish. Under this regimen, the embryos simply don't implant if the female hasn't fattened sufficiently for hibernation, thus ensuring her survival to breed another year.

As the rich foods of late summer and early autumn mature, bears accelerate the pace of their feeding in a race to accumulate sufficient fat to see them through the winter. Now, in addition to the usual plant foods, insects, and small rodents, the bruins eagerly seek foods such as acorns, beechnuts, apples, and corn. At this season of the year they can be especially destructive to farmers and orchardists. For that reason, bear hunting seasons often open in time to reduce or eliminate some of the damage caused by bears, especially sows with yearling cubs. As one biologist explained it, "A family of bears in a farmer's cornfield makes it look as if someone had driven a Mack truck around and around in the corn."

As winter draws near, black bears seek a den for hibernation. Although bears are commonly thought to dwell in caves and hibernate there, they don't need or use shelter during the warm months, and there are far too few caves to meet the bears' winter needs. Most black bears in northern regions hibernate beneath the roots of a blown-down tree, in a hollow in the ground, in a huge hollow tree, in a fox or coyote den that they've enlarged, or in some similarly sheltered place. In areas where summer cottages are abundant, such as parts of New York's Catskill Mountains, bears often den beneath porches or in crawl spaces under the uninhabited buildings. The human occupants leave well before the bears den in late autumn, and the bears depart in the spring before the humans return. It's a fine situation for the bears.

Where cold isn't extreme, black bears even build huge nests for winter quarters. These are nothing like the so-called "bears' nests" in beech trees; instead, they're constructed on the ground—great, dish-shaped circles of grass and leaves not unlike huge birds' nests. Once this edifice is complete, the bear curls up in it for the winter. Still farther south, bears may become torpid for a time, but don't have a winter den.

Denning time varies greatly, depending on the food supply rather than on the weather. In years when food is scarce, they'll den quite early. When good food is abundant, however, bears will keep on foraging for several more weeks, going into hibernation only when driven to it by severe winter weather.

In January or February the mother black bear gives birth to tiny cubs, usually two or three, which weigh only one-half to three-quarters of a pound. As with kittens, their eyes are closed at first and don't open for nearly a month.

A major misunderstanding is the belief that the female bear sleeps through the birth of the cubs, nurses them in her sleep, and wakes up in the spring to learn for the first time that—wonder of wonders—she has babies. The truth is that bears den, but don't hibernate in the strict technical sense of that word. In true hibernators, both heartbeat and body temperature drop dramatically; the temperature of denning bears doesn't drop during what we normally refer to as their hibernation period.

All of this means that bears don't go into a comatose state of true hibernation similar to that of the woodchuck. Instead they awaken frequently, move about in the den, and occasionally even emerge for short periods. In any event, the female is very much aware of her cubs, and takes good care of them in the den. She cuts their umbilical cords, nurses them, and is careful not to roll on them and crush them.

Despite the fact that they aren't true hibernators, bears have a complex assortment of wondrous adaptations to see them through the winter. For example, a bear's heartbeat slows to eight beats per minute for much of the time while it's denning. Moreover, it doesn't drink during that time, and "eats" only by consuming its own fat. Meanwhile, it neither urinates nor defecates until it leaves the den in the spring.

The cubs grow rapidly, for their mother's milk is extremely rich. To put this richness in perspective, consider that a black bear's milk contains at least 24 percent fat, while cows and humans produce milk with roughly 4.5 percent fat. It's no wonder that cubs thrive on such high-energy fare while still restricted to their den.

A thin mother and usually lively, well-fed youngsters emerge from their den in early spring. Black bears only breed every other year, so the cubs remain with their mother for a year and a half, learning how to fend for themselves. Finally, during the summer of their second year, the cubs set out on their own.

Mortality is very high among cubs during their first spring and summer—usually 50 percent or more. In areas where food is apt to be scarce in the spring, cubs often starve, but it also appears that other bears, especially males, kill many cubs. In one Arizona study, older bears killed half of the cubs that died from identifiable causes. In such cases the older bear also frequently eats the cub.

It's generally assumed that older, larger females will fight off males in order to protect their cubs, but that small, young, and inexperienced mothers aren't up to the task of driving off a much bigger male. This behavior is very difficult to document, however, since black bears are so elusive and their actions are generally concealed by forest growth.

Why would bears kill cubs of their own species? There are two plausible—though unproven—reasons. The first is that male bears can somehow identify their own offspring and kill only the cubs fathered by other males. This makes evolutionary sense, since the male is helping to ensure that, by eliminating a rival's cubs, its own genes will be those passed down through the population.

The second possible reason is that a female bear comes into heat within forty-eight hours after she stops nursing, following the loss of her cubs. The act of killing cubs thus presents the male with a double benefit: he has eliminated a rival's cubs, and he can now impregnate the female himself.

Black bears are extremely wary and difficult to see in the wild. Their rather poor eyesight is compensated for by marvelous hearing and sense of smell, and they use those faculties to avoid humans whenever possible. Consequently, many people who have spent a great deal of time in the woods, even in good bear habitat, have never seen a single truly wild bear. Our family has been extremely fortunate in this regard, for we've seen several bears on or very close to our land, as well as two thoroughly wild bears elsewhere. Two of these incidents were especially interesting.

Two years ago our daughter-in-law, Ellen, accompanied by our grandson, Davey, who was nine at the time, was driving up the road that leads to their house and ours. It's a very rural, winding road that goes uphill through the woods, with sharp banks on both the downhill and uphill sides.

As the car rounded a corner, there, sitting in the middle of the road, was a bear with two cubs! As Ellen and Davey exclaimed in astonishment at the sight, Mama Bear rose to her feet and, accompanied by the cubs, ran up the bank on the upper side of the road and disappeared in a flash.

While Ellen and Davey were still talking excitedly about what they had just seen, a *third* cub suddenly popped over the edge of the bank on the lower side of the road and emerged into the road. Instead of finding Mama there, as it expected, the little cub found only a big, frightening piece of machinery. There was only one thing to do, and that was seek safety by climbing a tree. Up the nearest big tree went the cub, where it began to wail piteously for its mother. After two or three minutes of the cub's outcry, Ellen wisely decided to leave, so that the mother could retrieve her errant cub. Sure enough, when Ellen and Davey drove back to the spot a few minutes later, the third cub was gone.

The most memorable of my own bear sightings took place on our land during the archery season for deer. I have a favorite tree stand that overlooks a rather open area with three apple trees often visited by deer. This happens to be the same tree stand from which I watched the porcupine plowing its way through the deep autumn leaves (see chapter 5).

It had rained earlier in the day, so I made little noise on the wet leaves as I approached the stand. Because I've occasionally arrived to find deer already feeding under the apple trees, I moved slowly and cautiously over a knoll that overlooks the area. There were no deer, but there was a black object under the nearest apple tree, about thirty-five yards away. My first thought was that it was a black Labrador, but I almost instantly rejected that idea in favor of a very large porcupine. Then it dawned on me that porcupines have very small, neat ears set close to the head, while this animal had large, rounded ears that protruded well above the head. It's a *bear cub,* I thought in complete astonishment!

At about that time, the cub sat down—not on its haunches, but on its rump, hind legs stretched forward, exactly as small children do. The ground was strewn with yellow apples, and the cub reached out with its front paws to grasp one. Then, front paws together, it held the apple for a long moment as if contemplating its quality or anticipating the taste of it. Then it finally opened its mouth and ate the apple. This identical performance was repeated numerous times, with the cub each time holding an apple in its front paws and gazing at it for a few seconds before consuming it.

Fascinated as I was by this performance, I was also quite uneasy. If there was a cub, there was also likely to be a mother nearby, and I didn't want to run afoul of her. I didn't even have an arrow nocked on my bow—not that it would have done me much good, for, as deadly as a hunting arrow is, a charging bear would do me in before the arrow killed it.

No doubt it would have been the better part of wisdom for me to retreat very slowly over the knoll and leave the area, but I was so intrigued by this rare opportunity to observe a wild bear that I hated to depart. Instead, I glanced about surreptitiously and quite apprehensively. To my great relief, there was no large, furry, black object within sight, so I decided to remain.

After consuming a large number of apples, the cub just sat there for a minute or two, doing nothing. Then it got to its feet and began to walk straight toward me. Finally, just as I thought that it might walk right up to me, it turned sideways and sat down on its rump, hind legs extended just as before, only twenty-five feet from me! Seeing the cub at that distance was an incredible experience, although I must admit that I was growing more apprehensive by the moment concerning the mother's whereabouts.

The cub sat there for no more than a minute, then rose unhurriedly. It took three or four steps back in the direction from which it had come, and then suddenly broke into a run. Although the woods were rather quiet because of the rain, it was nonetheless astonishing that the cub, which probably weighed about seventy pounds, could run at that pace and make scarcely a sound.

I wondered why the cub had left in such a hurry. Could it have smelled me? There was scarcely a breath of air moving, and even that was coming from the side, rather than from me to the cub. Still, the slightest eddy in the air current could have taken my scent to the cub. Could the mother have called to the cub? A bear's hearing is far keener than a human's, so the cub could well have responded to a call that I couldn't detect. I'll never know what actually caused it to run, but I'll always have the memory of that amazing encounter with a bear.

Scientists who have spent a great deal of time observing both wild and captive black bears tend to believe that they rank high on the animal intelligence scale. They're frequently reluctant to say so, however, since intelligence—be it in animals or humans—is difficult to define and often highly controversial. However, if we assume as a working hypothesis that intelligence includes curiosity, the ability to solve problems, and quickness in adapting to new situations, black bears indeed seem to be very intelligent creatures.

For example, one scientist who closely studied captive black bears says that they manipulate various objects, such as blocks, chains, and other playthings, at about the same level as chimpanzees. This would seem to indicate a substantial degree of curiosity about their surroundings.

Bears can also be remarkably creative about figuring out how to get at food. One park biologist told me about the case of a tightrope-walking black bear. In order to keep bears out of their food, campers in the park suspended their edibles in containers hung from a quarter-inch cable strung between trees. Park personnel were polite but privately disbelieving when campers came to them and said that they had seen a bear tightrope-walking the cable in order to reach the food containers. Skepticism turned to belief, however, when the campers returned with photographs of the bear caught in the act! In yet another instance, biologists observed a small bear standing on the back of a large bear in order to reach a sack of food suspended from a cable.

Pennsylvania biologist Gary Alt, who is one of the leading authorities on black bears, has witnessed some rather astonishing performances on their part. He has done a great deal of work live-trapping bears to weigh, tag, and radio-collar them. A suitably large trap, made from a section of culvert, has a piece of bait suspended toward the rear of the trap. When a bear enters and seizes the bait, a raised, hinged gate is released; this falls, latches securely, and imprisons the bear. Alt has known some bears, *after being trapped only once,* to return and stand with one hind foot raised while they take the bait, thereby preventing the falling gate from latching. Then the bears, bait and all, simply back out of the trap. If this isn't intelligence, it's at least remarkably rapid learning!

Alt also relates the astonishing tale of a 560-pound male bear that seemed to vanish into thin air. When Alt tracked this bear in the snow, he suddenly came to the end of its trail, with nothing around to climb or hide in. The first time this happened, he was understandably baffled by such an eerie sight, but he managed to solve the riddle after careful inspection of the tracks.

Without turning its hind feet, the bear twisted its body to the rear, jumped, and landed precisely in its own footsteps. Then, carefully stepping only in the tracks on its back trail, it traveled for a short distance before leaping off at right angles to the trail. Only by noticing that the tracks had claw marks at both ends was Alt able to figure out what the bear had done. What's more, Alt tracked the bear twenty-six times in two days, and the bear repeated this amazing trick every single time, backtracking with exquisite care from fifty feet to two hundred yards before leaping sideways off the trail!

Although intelligence is an extraordinarily difficult and contentious subject, it would appear, based on these and similar incidents, that black bears exhibit a great deal of behavior that most people would regard as intelligent. In other words, the black bear probably *is* the brightest bear in the woods.

The Brown Bear

Any consideration of North American brown bears must take into account the major differences between two very distinctive segments of this species—the inland grizzly and the coastal brown bear. Indeed, the differences between these two groups are sufficiently large that the two were long regarded as separate species. The coastal brown bear, a.k.a. the Alaskan brown bear or Kodiak bear, was listed as *Ursus middendorffi,* while the inland grizzly bear was *Ursus horribilis.* Now, however, most scientists agree that the inland grizzly and the coastal brownie are subspecies of the same species, *Ursus arctos.* Thus the brown bear is now *Ursus arctos middendorffi,* while the grizzly is *Ursus arctos horribilis.*

Inland grizzlies inhabit parts of the Northwest Territories, the Yukon, most of Alaska, British Columbia, and part of Alberta. In the lower forty-eight states, where they're an endangered species, the remaining grizzlies live mostly in and around two national parks, Glacier and Yellowstone.

Temperament is a good place to begin with these two-bears-in-one. Grizzlies are properly regarded as our most dangerous North American bear. *Ursus horribilis* it was named, Latin for "horrible bear," and its present subspecies name retains this view of the grizzly. Indeed, there's good reason for such an appalling name; there have been a great many recorded incidents of grizzly attacks on humans, with more than a few resulting in human fatalities. Grizzlies are highly protective of what they regard as their personal space, which can be very large in some cases. With that attitude, combined with a very short fuse, grizzlies should be considered extremely hazardous and strictly avoided.

Several years ago an experienced outdoor photographer was killed while taking a movie of a grizzly from what would seem to have been a prudent distance. When the bear suddenly charged, the photographer kept his camera running, and a major sporting magazine later showed horrifying still frames of the event, as the bear covered the distance to the photographer in a matter of seconds. The bear filled the last frames, as I recall, horrifyingly close. Tragically, those were the final pictures ever taken by the unlucky photographer.

In a more recent example, three hikers were mauled by a female grizzly with her cub. The first two unexpectedly happened upon the bears, and the mother immediately attacked them. One of the injured hikers staggered back along the trail to seek help from a park ranger. When he encountered another hiker, that person continued forward to help the injured person left behind. The bears were still in the vicinity, and the female attacked and mauled the would-be rescuer. Fortunately, all three hikers survived, although they suffered serious injuries.

These events should serve as a warning to others that they should stay as far as possible from grizzlies. At the same time, however, it's easy to exaggerate the dangers posed by grizzlies. The actual number of grizzly attacks on humans annually is extremely low, especially now that the Park Service has altered its policies on things such as storage and removal of garbage at campgrounds. To put the matter of grizzly attacks in perspective, a hundred or more people die annually from reaction to bee stings, to say nothing of the many thousands who fall victim to that great predator, the automobile. Grizzlies should be treated with the utmost caution and avoided wherever possible, but the chances that a reasonably cautious person will be attacked by one of these great predators is extremely small, even in prime grizzly habitat.

In contrast, coastal brown bears have a considerably more relaxed attitude toward people. That most emphatically doesn't mean that they're safe to approach closely, but these huge bears are far less likely than grizzlies to charge people at a distance. There are sites in Alaska, closely monitored to prevent people from straying too close, where the public can observe and photograph enormous brownies fishing for salmon. These sites are located at distances from the bears that would be far too short to contemplate with inland grizzlies.

While black bears are adept at climbing trees, brown bears, whether inland grizzly or coastal brownie, lack that ability. Thus trees, if available, can sometimes offer safety from bear attacks, and more than one person has survived a grizzly attack by seeking refuge in a tree.

Size also sharply differentiates the two divisions of the brown bear population. On average, grizzlies are substantially larger than black bears, but much smaller than coastal brownies: adult grizzlies weigh from three hundred to nine hundred pounds or more, but some of the giant coastal brown bears exceed 1,500 pounds.

This major difference in size is attributed in large measure to diet: while grizzlies eat substantially more meat than black bears, coastal brown bears have a far more meat-rich diet than either. Grizzlies prey on deer fawns and elk calves, and also dig a variety of small rodents out of the ground. At times, they also fatten on huge hatches of army cutworm moths, which swarm under rocks, or devour large groups of ladybird beetles. Still, grizzlies lack a steady supply of meat.

Coastal brown bears, on the other hand, have access to most of the same food sources available to the inland grizzly, plus several other incredibly rich supplies of meat. In particular, there's the annual spawning run of salmon, but the ocean also casts up such savories as dead or injured seals and whales. In sum, the coastal brownies have access to an abundance undreamed of by grizzlies.

Brown bears, like black bears, usually give birth to two or three cubs in a litter, but, unlike black bears, they only breed every third year, so that the cubs remain with their mother for two and a half years. The cubs are born in a den, which in the case of most grizzlies is at a rather high altitude—six thousand to seven thousand feet. Often the bears create dens by digging into steep slopes. They also excavate beneath huge logs or utilize caves if they're available. Coastal brown bears, for obvious reasons, den at much lower altitudes, but their denning sites and habits are otherwise like those of grizzlies.

Another difference between the two subspecies is color. As its common name indicates, the coastal brown bear varies from light to medium brown. The grizzly, on the other hand, is darker and has white tips on the hairs of its back, in particular. This intermingling of dark and white presents the grizzled aspect from which the bear gets its name.

The coastal brown bears, and the inland grizzlies of Alaska and Canada, seem to be holding their own in most areas. Grizzlies in the lower forty-eight states are another matter. Relentlessly persecuted for years, like the wolves, grizzlies have been reduced to a small fraction of their former numbers and range in the American West.

As previously mentioned, the only two major concentrations of inland grizzlies in the lower forty-eight states are in and around the Yellowstone and Glacier National Park systems. Now, however, the U.S. Fish and Wildlife Service hopes to reintroduce grizzlies into the Selway-Bitterroot area along the Idaho-Montana border, where the last of their kind were exterminated over fifty years ago.

Brown (grizzly) bear

By reintroducing twenty-five to thirty grizzlies over the next five years, biologists hope eventually to see as many as three hundred of the big bears inhabiting the area. With 5,700 square miles of designated wilderness on public land in this region, it's the only remaining area in the lower forty-eight states capable of supporting such a restoration. As with so many other endangered species, the fate of the grizzly rests on the maintenance of adequate habitat.

The Polar Bear

The extraordinarily barren, frigid world of the polar bear *(Thalarctos maritimus)* is nothing like that of black or brown bears. It follows, then, that the polar bear is a vastly different bear from its North American relatives—indeed, from all other bears. Living in such a harsh environment requires the polar bear to have numerous adaptations, some of which are truly amazing!

The polar bear's species name, *maritimus,* correctly designates it as the seagoing bear, a creature at home not only along the ocean but in it, so its adaptations begin with this central fact of its existence. Polar bears float easily and are powerful, almost tireless swimmers. Individuals have been sighted by ships, miles from shore and swimming with apparent ease, in water so frigid that most creatures, including humans, would perish in it within minutes. It even appears that these remarkable seafarers can sleep while at sea!

The reason for this amazing endurance resides in the bear's underfur. This coat is so dense that it traps a great deal of air, which lends its owner the buoyancy needed to swim long distances with such ease. Perhaps even more important, it serves to keep water away from the bear's skin, thus effectively insulating it from the icy ocean. Added to buoyancy and insulation, the polar bear's huge front paws, edged with stiff hairs to make them even wider, act as great paddles to propel the owner easily through its aquatic environment.

Not surprisingly, polar bears subsist almost entirely on meat, for plant materials are scarce in this great bear's habitat. Berries and other plant foods are consumed when available, but the seal is the polar bear's main food supply. Such things as nesting birds and their eggs, as well as stranded whales, supplement the bear's diet.

Polar bears can be incredibly patient hunters, waiting for hours beside a seal's breathing hole to dispatch the seal when it comes up for air. They will also stalk seals on the ice by doing what one scientist terms "otter travel," in reference to the otter's rather humpy land gait, which tends to emphasize the height of the hindquarters. In this mode, the bear flattens its head and front quarters against the ice, tucks its front paws beneath it, and propels itself slowly toward the seal with its hind feet and slightly elevated rump. The sight of this huge beast creeping forward in such a strange position must indeed be an astonishing one!

In addition to their other remarkable aquatic skills, polar bears can also dive to a depth of up to fifteen feet. They sometimes use this ability to pursue an escaping seal, or to swim underwater to reach a seal basking on an ice floe: at the edge of the floe, the bear hurls itself out of the water to seize its prey.

The coastal brown bear is the world's largest bear, right? Not so fast—some knowledgeable scientists aren't so sure. They think that the polar bear at least equals the brown bear in size, and may occasionally exceed 1,600 pounds. Understandably, reliable polar bear weights are hard to come by; given the remoteness of most polar bears, weighing them, alive or dead, can be a daunting proposition. For the present, then, we can think of the brown and polar bears as Their Ursine Majesties, sitting atop thrones of equal height as the planet's largest land carnivores.

Almost everyone knows that polar bears are white—but almost everyone is wrong! The polar bear's white appearance is part of an astounding twin adaptation that allows the bear to survive and function in a land of almost unimaginable cold. The first adaptation is in the bear's long outer hairs. The center of each hair conducts light, much like a fiberoptic strand. Some of this light filters through the translucent outer part of the hair, making it *appear* white, although it doesn't have pigment to make it truly white. The second adaptation is black skin, which readily absorbs the heat from the sunlight filtering down through the centers of the hairs. These adaptations are highly unusual, although not unique, for arctic foxes and seal pups also share them.

These adaptations combine to give the polar bear the best of two worlds, as it were. Without the white-appearing coat's camouflage, it would be extremely difficult for the great predator to stalk and kill its prey in a largely white world. Also, lacking the black skin underneath to absorb the heat of the Arctic sun during that part of the year when it's visible, the bear would require a great deal more energy to keep itself warm. Working in tandem, these wonderful adaptations help ensure the polar bear's success in its forbidding environment.

Although polar bears have been known to attack people, and certainly should be considered dangerous, scientists believe that they're far less aggressive and temperamental than inland grizzlies. They're certainly not afraid of humans and may approach out of curiosity, with no aggressive intent; some have likely never laid eyes on one of our species before, which might also be a cause of their curiosity. Some polar bears, however, are highly aggressive, so all of them should be regarded as armed and dangerous. As proof of this, an Inuit campsite on the western shore of Hudson Bay was recently attacked by a polar bear. A woman was killed, and a man and boy were seriously injured.

Like the brown bear, the polar bear breeds only every third year. It usually has two cubs, though occasionally only one. The female dens before giving birth, although the word *den* can be something of a misnomer in the polar bear's case. Often the female will dig a den in a snowbank or a peat bed, but sometimes she'll simply curl up in the snow and allow falling and drifting snow to cover her deeply enough so that a small but snug cavity is formed. Remarkably, males, as well as females with cubs, don't den, but remain active through the incredibly bitter arctic winter!

Because polar bears, unlike their black and brown cousins, are completely exposed to view, attacks by male polar bears on cubs, as well as the mothers' defense of them, are well documented. According to Gary Alt, their method of attack is vastly different from that of brown and black bears. In the case of the latter two, the attack is a savage rush that is quickly over. Either the female successfully defends her cubs against the oncoming male, or he kills one or more of them. With polar bears, since the female and cubs are visible at all times and have no place to hide, the process is a long, slow stalk covering mile after mile.

While the male stalks behind, the female keeps a watchful eye on him as the cubs hold tight to her flanks. Every so often, when she loses patience, the female turns and charges the male. Amazingly, the cubs charge with her in a move that Alt says is so beautifully choreographed that it's as if only one body were moving, instead of three!

Only rarely does a cub separate from its mother, in which case it's usually seized and killed by the male. More commonly, however, the female, with cubs glued to her sides, charges the male every little while, persevering until the male at last gives up and goes away.

On the whole, our three North American bears seem to be thriving. No longer viewed as nuisance animals, or worse, to be eradicated as quickly as possible, they're now being seen for what they are—fascinating, immensely powerful predators that have a place in the scheme of things. True, grizzlies in the American West still have many detractors and are controversial, but a rapidly growing number of supporters seek to retain them or reintroduce them where possible. As large, free-roaming predators, all three species depend on major blocks of suitable habitat as the key to permanent survival. As long as we protect and maintain such habitat, these magnificent animals will continue to be a part of our world.

Polar bear

22

The Comeback Kid: The White-tailed Deer

Myths

- Deer have horns.
- A deer's age can be told from its antlers.
- The deer's breeding season is brought on by cold weather.
- Newborn fawns have no scent.
- A fawn found alone is probably orphaned.
- A doe will abandon her fawn if humans have touched it.
- Deer make good pets.
- Deer are quite tall.
- Deer are fragile and easily injured.
- A winter deer yard is much like a woodland barnyard.
- Deer live many years in the wild.
- Small dogs pose little threat to deer.
- When the snow becomes deep, it's time to start feeding deer.

If a poll were taken to determine North America's single most important mammal, it would surely be the white-tailed deer (*Odocoileus virginianus*). Nearly ubiquitous and highly adaptable, it can be found in almost every imaginable habitat: farmland, brush, deep forest, desert, swamp, suburb, and sometimes even in rather densely populated portions of cities throughout most of the lower forty-eight states and the southern third of Canada. Moreover, deer are abundant, often excessively so, throughout most of their range, so that they've become a familiar part of the daily lives of countless millions of people.

White-tailed deer are indisputably our most economically important wild mammal. As our premier big game animal, deer generate hundreds of millions of dollars in retail sales annually, and billions of dollars in overall economic impact. In addition, there is the monetary value of some 250 *million* pounds of venison annually, which, at a conservative value of two dollars per pound, is worth $500 million, to say nothing of the intangible value to countless hunters and their families of putting a highly prized meat on the table through their own efforts.

Equally important is the intangible value of seeing and watching deer. Graceful, elegant, and beautiful, whitetails command the admiration and affection of large portions of our continent's human population, who love to watch these marvelous creatures. They add immeasurably to our quality of life, and their presence serves to remind us of the values of a natural world often forgotten in the frantic pace of our daily lives. By almost any human measure, the white-tailed deer is the undisputed favorite of our North American mammals.

The origin of the whitetail's common name is obvious to anyone who has seen the animal when it's suspicious, nervous, or in flight. When a whitetail is unconcerned, its rather long tail points downward and is quite inconspicuous; the exposed upper side is mostly brown, with a narrow black border and just a touch of white showing at each edge.

At the first sign of danger, however, a dramatic change occurs. The tail shoots up in an erect position, revealing an all-white underside that gleams like a beacon. To make this warning signal even more impressive, the deer flares the long white hairs out to the side and wags this highly utilitarian appendage from side to side as a warning signal. When displayed in this fashion, it's easy to understand why a white-tailed deer's tail is called its "flag."

In sharp contrast with its common name, the whitetail's scientific name is something of a misnomer. Its genus name, *Odocoileus,* is probably a misspelling of *odontocoileus,* Greek for "hollow tooth." This name is the work of an early French-American naturalist with the rather picturesque name of Constantine Samuel Rafinesque, based on a single tooth that he found in Virginia. Whitetail teeth aren't normally hollow, and Richard Nelson, in his splendid book *Heart and Blood: Living with Deer in America,* speculates that Rafinesque might have found a thoroughly atypical hollowed-out tooth of a very old deer.

Rafinesque himself is worth a small digression, for he was truly a bizarre and astonishing character, even more picturesque than his name. His brilliant intellect, which was formidable, was exceeded only by his eccentricity and arrogance. A zealous and shameless self-promoter, he made enemies even of those who greatly admired his intelligence and boundless energy. More to the point, his zeal for identifying new species sometimes led to hasty judgments that marred his otherwise admirable contributions to natural history. Such appears to have been the case with the scientific name he attached to the white-tailed deer.

The species name, *virginianus,* conferred by Rafinesque because the tooth was found in Virginia, also seems unduly provincial for a creature as far-ranging as the whitetail. However, it at least has the virtue of locating the approximate center, from north to south, of the animal's range in North America, although subspecies of the whitetail range down through Central America to areas below the equator in South America. If any proof were needed of the whitetail's phenomenal adaptability, this enormous distribution is it!

Unlike many other North American mammals, which originated in Asia or South America and later migrated here, white-tailed deer evidently evolved on this continent roughly 4 million years ago. Their ultimate forebears, primitive hoofed mammals, or ungulates, which deer indisputably are, date back to the Paleocene Epoch, 65 million to 55 million years ago. This was just after (in terms of the unimaginably long ticks of the geologic clock) the extinction of the dinosaurs, so our modern deer have indeed had a lengthy evolutionary journey!

The numbers of deer living in America during pre-Columbian times is a matter of considerable debate. Some estimates, based on the accounts of early explorers and settlers, run as high as 50 million. Others believe that these early accounts were exaggerated or misinterpreted, and that we now have more deer than were present before the arrival of European settlers.

Certainly the present mix of habitat—forest, brush, and fields—seems more conducive to high deer populations than the forests primeval. But those who support the high estimates believe that the enormous quantity of mast-bearing trees—oaks, chestnuts, and beeches—provided a rich enough source of autumn food to compensate for the relative lack of browse in these virgin forests.

We'll never know for certain how many deer were here in pre-Columbian days; it's perhaps enough to know that they were very abundant. What we do know, however, is that this bountiful resource was nearly squandered because of such things as ignorance, carelessness, and greed, and that whitetails have now rebounded to an estimated 25 million to 27 million animals.

The two most important reasons for the whitetail's precipitous decline were habitat loss and unregulated hunting. As the forests were systematically leveled for pastures, cropland, timber, and firewood, good deer habitat became both scarce and fragmented in many parts of North America. My own state of Vermont, for instance, was 70 percent cleared just prior to the Civil War.

Deer were highly valued for their excellent meat, as well as for their hides and tallow. Unfortunately, there were no notions of modern game management principles at that time: there were no bag limits, and deer could be hunted year-round by virtually any method. Some insight into this state of affairs was given by the nineteenth-century Vermont author Rowland Robinson.

A Quaker whose home was an important station on the Underground Railroad, Robinson was also a gifted writer and painter, as well as an avid hunter and early conservationist. In one of his novels, the principal character inveighs against the practice of "crustin'." This method of hunting consisted of waiting for a crusty snow that wouldn't hold a deer's weight but would sustain the weight of a man on snowshoes. A hunter merely had to locate a deer and go after it; the deer, quickly exhausted by breaking through the crust and floundering in deep snow, was easy prey for the hunter on snowshoes.

By the mid-1800s, the seemingly limitless supply of white-tailed deer in the United States had dwindled to an estimated 1 million animals, a pathetic remnant of a once-great population. Fortunately this was the whitetail's nadir, because alarmed sportsmen conservationists began to press for measures to protect and restore this magnificent animal.

These measures took several forms. First, laws were passed in some states banning all deer hunting. Second, whitetails were live-trapped in areas where they were still present in moderate numbers, and then used to restock areas where deer were either extremely scarce or had been eliminated. Vermont again serves as an example. Deer hunting was outlawed in 1865 and remained so for thirty years. Subsequently, a number of deer were imported from New York to restock portions of the state where deer had been nearly eliminated.

As helpful as these steps were, habitat regeneration was even more important. It's axiomatic in wildlife management circles that there can be no wildlife abundance without decent habitat, and resurgent whitetail populations proved it. Agriculture was rapidly evolving and becoming more mechanized; this meant that the roughest little hardscrabble hill farms were abandoned as farming became concentrated on the better lands. Also, sheep raising and other forms of agriculture moved westward after the Civil War, triggering more farm abandonment. Abandoned agricultural lands reverted to brush in a few short years, thereby creating prime habitat that the now-protected deer speedily filled.

One major reason for the whitetail's rapid salvation from the brink of the abyss is its reproductive rate, which is very high for a large mammal. When deer are well nourished, many does only six months old will breed and bear a fawn the following spring. Older does generally have twins, triplets are by no means rare, and quadruplets and quintuplets have been reported on a few occasions. This reproductive potential, if unchecked by predation, can produce an astonishing number of deer in a very few years.

In one experiment, an area of roughly one thousand acres was surrounded by deer-proof fence. Biologists then used controlled hunting to cut the number of deer in this enclosure to ten. Then they stopped all hunting to see what would happen. What happened was that the ten remaining deer multiplied to 212 in only five years!

By the turn of the century, whitetail populations had rebounded to huntable levels in many areas of former scarcity, and strictly regulated hunting seasons were established. At this point another piece of the jigsaw puzzle of whitetail biology came into play: the mating habits of the species. This, as we shall see, had major consequences that were to prove both a blessing and a curse to wildlife managers and the deer themselves.

To understand this part of the puzzle, it's necessary to digress and explore the nature of the male deer—the whitetail buck. First, two widely held notions about the whitetail buck need to be corrected. One is that they have horns. Wrong! Cattle, sheep, and goats have horns, but not members of the deer family. Horns are hollow, permanently attached, and grow throughout the owner's life. Deer, on the other hand, have antlers. Composed of exceedingly hard bone, they're shed each year during the winter, and the buck grows a replacement set.

Antler development begins in early spring. The developing antlers, starting with tiny nubs, are full of blood vessels and nerves, and have a brown covering called "velvet" because of its resemblance to the fabric. Bucks at this stage, which lasts most of the summer, are said to be "in the velvet."

In September or October, depending on latitude, bucks rub their antlers against saplings to remove the velvet and polish them. The structure of these antlers is unique. When fully developed, they consist of a curving, approximately horizontal main beam with *single* sharp tines projecting vertically at intervals (in contrast, mule deer tines rise from the main beam and then fork). A buck's antlers, in toto, are termed its "rack," and the tines, including the tips of the main beams, are called "points." Thus a buck with four points on each side is said to have an "eight-point rack," or is simply called an "eight-pointer."

The second long-standing misconception about deer is that a buck's age can be determined by its rack. According to this theory, a buck grows spikes (two points) in its first year, four in its second year, six in its third, and so on. While this is more or less true under some circumstances, it's wildly inaccurate in others.

Although older, mature bucks do tend to have larger, heavier racks than young bucks, antler development is largely a function of nutrition and genetics. Thus well-fed yearling bucks with the right genes often have six or eight points in some areas, while a poorly nourished two-year-old, or one with inferior genetics, may grow only spikes. The only accurate way to determine the age of a deer is by its teeth. For the first two years, development and replacement of teeth are a reliable indicator. Thereafter, tooth wear serves as a guide, although it becomes increasingly less accurate as the deer ages. In older deer, the only truly accurate way of assessing age is by examining a tooth's annual growth layers under a microscope.

With the advent of autumn, the breeding season or "rut" begins. Another myth, incidentally, is that cold weather triggers the rut. Actually, biologists have learned that waning light intensity is the cause; cold weather is simply coincidental. The rut starts in late October and peaks in November in northern climes, but may be a month or more later in more southerly latitudes.

Then those assiduously polished antlers come into play, as bucks vie with each other for dominance. Nonterritorial for most of the year, bucks now try to establish hegemony over a territory that may embrace hundreds of acres. Often dominance is established simply by minor sparring matches that quickly convince younger, smaller bucks that discretion is the better part of valor. More evenly matched bucks, however, may engage in genuine battles, locking antlers and attempting to shove each other around so ferociously that the ground is torn up and shrubs trampled.

Eventually, one buck performs the whitetail equivalent of "crying uncle" and turns to flee, sometimes being pursued and gored in the process. On very rare occasions, bucks are unable to unlock their antlers; bound together like macabre Siamese twins, the two finally perish from starvation!

Now comes the piece of deer biology that has been both blessing and curse: the highly polygamous nature of whitetail bucks. Once a buck has established dominance in his territory, he proceeds to breed as many does as possible. Aware of this fact, wildlife professionals utilized it when they began to reinstate hunting seasons for rebounding deer populations. By limiting the seasons only to antlered bucks, they left the population's doe segment untouched. Since one buck can breed many does, this allowed deer numbers to continue their climb unchecked. For many years, this was the blessing. Then came the curse.

"Buck only" seasons became the rule in most of the major deer hunting states. They worked extremely well—far *too* well in the long run. With wolves and cougars—the traditional major predators of whitetails—eliminated, humans exercised the only effective control over deer numbers. Deer populations climbed and climbed for decades. For a while, regenerating habitat kept up with this increase, but finally overpopulation began to overwhelm their food supply.

The answer, to wildlife biologists, was obvious: begin hunting female deer in order to limit the population. So one state after another began to extend hunting to deer without antlers—popularly called "doe seasons." Thus were born what have been called "the deer wars," bitter battles between deer hunters and wildlife biologists.

Several generations of deer hunters, as well as nonhunters who simply enjoyed seeing deer everywhere, had watched deer numbers soar under buck-only seasons, and it had become a deep-seated article of faith that killing female deer would destroy a deer population. Now they were being told that this belief was wrong, that deer couldn't be allowed to increase forever, and that for the good of the habitat and the future of the deer, the population increase had to be halted.

The majority of them didn't buy it. In state after state, the debate became extraordinarily rancorous. Wildlife biologists were derided as "college boys who don't know anything that didn't come out of a textbook"—and were often called a great deal worse. There were angry mass meetings that had many of the trappings of a lynch-mob mentality, and indeed, at least one prominent deer biologist was hanged in effigy!

Nonetheless, by fits and starts, one state after another initiated antlerless deer seasons to keep deer in balance with their habitat. Despite the predictions of disaster from opponents, this system of management soon proved itself. Deer were larger and healthier, while deer populations remained at a relatively high, yet stable, level. In time, the screaming opposition to deer management first died to a dull roar, then to an undercurrent of muttering, and finally changed to support by all but a diehard cadre of hunters and deer watchers.

At the risk of seeming chauvinistic, I'd again like to cite Vermont as something of a microcosm of the deer wars. Progress came late to Vermont, and it was the very last state to adopt regular antlerless deer seasons. By the 1960s, the critical winter food supply was terribly overbrowsed. According to every measure of deer health—weight, antler development, and reproductive rate—Vermont's deer were in abysmal and steadily deteriorating physical condition.

Despite the resounding success of other states' deer management programs, there was fierce opposition to any antlerless deer hunting from sportsmen and dedicated deer watchers, who could drive around on a summer evening and count a hundred or more deer grazing in fields.

The legislature, over strong protests, finally allowed the Vermont Fish and Wildlife Department to hold a few sporadic and severely limited antlerless deer seasons, but these were too little and too late. At an estimated 250,000—nearly twice what the land could sustain on a permanent basis—the state's deer herd was a disaster in the making, and supporters of the status quo were living in a fool's paradise.

The crash, which wildlife biologists had predicted for a number of years, finally arrived in spectacular fashion. During the severe winters of 1969 and 1970, an estimated 100,000 deer died, most of them slowly and painfully from the effects of malnutrition. No matter—the crash was blamed on the limited antlerless deer seasons which had been held, and the legislature hastily returned the state to bucks-only seasons.

This was precisely the wrong approach, and after nearly another decade, the deer population actually declined still further. Finally, in response to harsh criticism, legislators tumbled over one another in their eagerness to toss this political hot potato to the Fish and Wildlife Department.

What followed was a bitter pill, but one that had to be swallowed. The department initiated a series of seasons designed to cut the deer population well *below* the normal carrying capacity of the winter range and hold it there, thus allowing winter food supplies to regenerate.

Predictably, this policy created a great deal of anger and criticism, but the department held firm. Then, once winter food supplies were again in good shape, the deer herd was allowed to increase slowly until it was at or near what the winter range could sustain. Now, as in other states, scientific deer management—including antlerless deer hunting—enjoys widespread support in Vermont.

Wolves and cougars were originally the predators that kept deer numbers in balance with their habitat. When those two predators were for all practical purposes eliminated, humans became the only remaining significant control over whitetail numbers. As already indicated, modern wildlife managers, by the use of controlled antlerless deer seasons, have largely been able to keep deer numbers at or just below the carrying capacity of their habitat.

There are a few exceptions, though. The major ones are suburbs, the fringes and less densely settled parts of urban areas, and places that are isolated either physically or by some special category of ownership. These latter include such things as islands and special reserves of one sort or another.

At this point the "Bambi Syndrome," dreaded by biologists and anyone else who believes in wildlife management based on science rather than emotion, comes into play. A brief digression is warranted to explain this phenomenon. The original *Bambi* was a book written in German by Felix Salten. Although anthropomorphic, it is a rather dark, philosophical allegory that in many ways accurately portrays the many hazards facing European roe deer in the wild.

Evidently intended as a book for adults, *Bambi* became popular with children, especially after its translation into English in 1928. Then Walt Disney, taking considerable liberties with the original tale, made *Bambi* into an immensely popular feature-length cartoon of monumentally anthropomorphic proportions. In his hands, wild creatures precisely fitted the description that I used to see on bumper stickers: "Animals are just little people in fur coats." The results of this are all too evident to this day, for in a recent Internet visit to the movie version of *Bambi,* I found an attached review by a sixteen-year-old boy who said that the movie teaches us how destructive hunting is.

To return to the problem of restricted areas, be they suburbs, islands, or special reserves, the subject of whitetail overpopulation and the control thereof is often highly contentious because of the Bambi Syndrome. Many inhabitants of these areas simply can't bear the thought of a deer being killed by humans—thereby ignoring the unpleasant facts of death by malnutrition, parasites, disease, and other forms of mortality that especially afflict overpopulated deer.

In such areas, the issue of deer control often pits neighbor against neighbor in rancorous disputes and strains the bonds of friendship to the breaking point. As gardens are ruined, expensive shrubbery consumed, and landscaping devastated by a hungry horde of deer, angry homeowners often liken them to outsized rats! The steady rise of expensive collisions between deer and automobiles further exacerbates the situation, and as Lyme disease (transmitted to humans by deer ticks) increases, excess numbers of whitetails become a health concern as well.

Eventually in such situations, pressure increases dramatically for control of the deer by those who are fed up with the problem. At that point other residents rush to defend the deer, and what has at heart been a biological problem now becomes a very sticky sociological one, with angry people on both sides of the issue.

Other than letting starvation and disease decimate the deer, there are really only three methods of dealing with whitetail overpopulation in limited areas: carefully regulated public hunting; hiring professional marksmen; and trapping live deer and transporting them elsewhere (a number of attempts at trapping or tranquilizing deer and sterilizing them with various hormones have proved both impractical and unsuccessful—besides costing about one thousand dollars per animal).

Killing deer as a population-control measure is almost sure to raise a ruckus in suburb, island, park, or reserve. As a result, people living in or near these locales, often well off financially, frequently choose the "trap and transfer" method. Ironically, this technique is apt to prove far more traumatic and cruel to the deer than shooting a portion of the population.

Many studies have been carried out to determine just what happens to deer that have been trapped and transported to new quarters, far from their original range. The results have been dismal, to say the least. First, about 20 percent of these deer initially perish just from trauma, although they may not die for several days after their release, thereby giving the impression that the operation was a great success.

Beyond that, these deer are now strangers in totally unfamiliar surroundings which already contain as many whitetails as the land can support. Car collisions, disease, and predation—human and otherwise—take a rapid toll. In the end, 40 percent to more than 80 percent of the transported deer are dead within a year.

Added to all this is the expense, which is high. It costs roughly four hundred dollars to trap and transport each deer, or forty thousand dollars to remove one hundred deer from an area. Moreover, this expenditure has to be repeated every year or two because of the whitetail's high reproductive rate.

Some communities have chosen to hire professional sharpshooters, who typically work at night from raised platforms adjacent to baited areas. Two of these are the 1,200-acre University of Wisconsin Arboretum and 1,200-acre Long Island in New Hampshire's Lake Winnepesaukee. Expenses for this option vary somewhat according to a variety of circumstances, but seventy-five to one hundred dollars per animal seems to be typical.

Finally, other places have taken the option of using public hunting as a means of control. Often the hunters are carefully selected and must pass a marksmanship test. For safety reasons, only relatively short-range weapons are used—shotguns, muzzleloading rifles, and bows. One of the prime attractions of this option is the fact that it's essentially accomplished with little or no public expense (hunters' license fees usually more than cover any operational expenses incurred by the state wildlife agency).

Examples of this approach encompass a variety of situations. The two-thousand-acre Mary Flagler Cary Arboretum in New York State has used controlled public hunting for nearly thirty years to keep deer in check. The land surrounding the big Quabbin Reservoir in Massachusetts—the water supply for Boston—has more recently utilized limited public hunting to reduce the number of deer, which were destroying any forest regeneration. On a larger scale, Connecticut has had considerable success with a special season in what it calls the Urban Corridor zone.

In the end, it all boils down to one thing: the deer are going to die anyway, so it's only a question of what method to use. Even those who advocate doing absolutely nothing and leaving the deer strictly alone in these sheltered enclaves of overpopulation are only choosing a different form of death. Left to their own devices, these increasingly overpopulated deer suffer from malnutrition and its associates—disease, parasites, and, if nothing else claims them first, ultimate starvation.

Returning to the whitetail's mating habits, a doe is usually bred by the area's dominant buck, though sometimes by a lesser buck who has the good fortune to be in the right place at the right time and, in a deer's version of Confederate General Nathan Bedford Forrest's reputed philosophy, "got there firstest with the mostest."

A gestation period of about two hundred days ensues. Because most does are bred during their first estrus, this means that the majority of fawns in northern latitudes are born in May or early June, but there are always some that aren't bred until their second or even third estrus, so that some fawns arrive as late as July and August. These late-born fawns are so small by the time cold weather and snow arrive that their chances of surviving the rigors of winter are very slim.

Whitetail fawns are among the loveliest and most endearing of all creatures. Weighing only six or seven pounds at birth, the tiny creatures come equipped with a reddish-brown coat and lines of white spots; the overall effect makes excellent camouflage for a fawn resting quietly on a sun-dappled carpet of dead leaves.

Perhaps it was inevitable that such appealing little animals would generate their own particular misconceptions, and there are at least three major fallacies surrounding fawns. The first is that fawns are virtually scentless as a defense against predators. According to John Ozoga, a consultant and former research biologist with the Michigan Department of Natural Resources, even newborn fawns have a definite scent, and research biologists have used dogs on occasion to locate fawns by scent.

It's true, however, that very young fawns have a scent that's much less pronounced than that of adult deer. In addition, very young fawns lie down and stay where their mother puts them, so that they don't leave a scent trail. Together, these two traits act as a defensive mechanism and clearly make it considerably more difficult for predators to locate a fawn; still, the fact remains that a fawn isn't completely scentless.

A second fallacy is that a fawn found by itself is either orphaned or abandoned by the mother. This erroneous idea has led to all kinds of problems relating to people taking "orphaned" fawns home to raise, in clear violation of the laws of most states. The fact is that a doe habitually stashes her very young offspring and wanders off to feed, returning two or three times a day to nurse them. Sometimes she's close by, but at other times she may be at a substantial distance. She always knows precisely where both she and her young are, and there is not the slightest chance that she will lose track of them.

Fawns are actually orphaned sometimes, of course. Only the does care for the fawns; as with most mammals, the buck assumes no family responsibilities whatsoever after breeding the doe, so the fawn is effectively orphaned if something happens to its mother. For instance, a doe may be killed by an automobile or by predators, or may die from other causes, thereby leaving an orphaned fawn.

A recent personal experience is illustrative. One day in May two years ago, I heard a peculiar and unfamiliar sound through an open window. Roughly halfway between a mew and a soft bleat, it continued for an hour or two, but finally subsided. It resumed the following morning while I was working in the garden just below our home, and kept on and on. The sound, sometimes almost inaudible and at other times fairly loud, seemed less than a hundred yards away and appeared to emanate from a spot just inside the woods.

Curiosity got the better of me, and I finally entered the woods to seek the source of this unusual noise. It took me some little time and considerable quiet sneaking about, because the sound had a ventriloquial quality, seemingly coming first from one spot and then from another. Finally, however, I saw an obviously very young fawn standing in an area of ferns that overtopped it by a considerable margin.

I backed quietly away, returned to the house, and called Gordon Marcelle, our local game warden. I told him that the fawn's prolonged bleating over two days struck me as abnormal, and that it seemed likely something had happened to its mother. He replied that fawns with healthy mothers had occasionally been known to bleat a great deal. He counseled patience and said that if the bleating continued the following day, he would come to pick up the fawn.

The bleating resumed the next day, I phoned the warden, and soon he and his wife appeared, carrying a baby bottle with a sugar-water mixture in it. Warden, wife, and I, followed by our two grandchildren, soon located the tiny fawn, sound asleep at the base of a tree. The warden examined the little creature, which offered no resistance, and said there was no evidence that the fawn had been fed, indicating that it was orphaned.

We all adjourned to our living room, where Mrs. Marcelle induced the fawn to take a little sugar water, and our grandchildren, aged eleven and nine at the time, were allowed to hold the tiny fawn. Meanwhile, I busily took photographs and my wife admired the beautiful little creature, whose tiny hooves were no larger than my thumbnail. The fawn, the warden told us, would be taken to a veterinarian who would care for it in such a manner that it could eventually be returned to the wild, rather than becoming a pet. This was an outcome that pleased us greatly.

This isn't the end of the story, however. Eight or ten days later I noticed two turkey vultures flying low near our house. The big scavengers have gradually worked their way up to northern Vermont, and we usually see them at low level a time or two each summer, so I attached no special significance to their presence. Even when the pair landed for a few minutes in an evergreen tree near where we found the fawn, I failed to comprehend the reason for their behavior. Then, three or four days later, the putrid odor of a large, decaying animal began to assail us when the wind blew from the woods toward our house. As the smell grew worse, I finally went in search of its source and, after a lot of decidedly unpleasant sniffing and searching, found the badly decayed remains of a rather small deer not fifty yards from where we had found the fawn.

It wasn't difficult to make some valid deductions. Almost certainly, this was the mother of the orphaned fawn. There was no sign that she had been attacked by predators, and a doe would lead attacking dogs or coyotes far away from her fawn anyway, even if they had been able to catch and kill her. From her small size, it seemed probable that she was barely a year old and the fawn was her first. Very likely something went wrong during birth, and the young doe bled to death at or very near the birth site. Such things happen more than we may realize in the natural world, although we're seldom aware of them.

Despite the occasional fawn that is truly orphaned, there's a very clear message regarding fawns found without a mother in sight: DON'T pick them up and take them home! Instead, call your local conservation officer or game warden. He or she can check on the situation and take appropriate action—although in most cases no action is required. Usually the doe is either feeding some distance away or is lurking in the brush just out of sight, waiting for the human intruder to leave so that she can retrieve her baby.

This leads to a third common fallacy about fawns—that once humans have handled one, its mother will abandon it. There is a small grain of truth in this notion, but only a *very* small one. Does have indeed been known to abandon fawns that have acquired a dose of human scent, but this is very uncommon. According to John Ozoga, who has handled hundreds of fawns, a fawn born so recently that it's still wet will retain more human scent than a dry one. Likewise, he says, a young doe with her first fawn is more likely to be skittish of human scent than an older, more experienced doe. Usually, it takes the combination of a yearling mother and handling a wet, newborn fawn to trigger abandonment.

More common are instances in which wardens or biologists have taken an "orphaned" fawn from someone who has picked it up, carried it back to the place where it was found, and then hidden nearby to await results. In most cases the doe showed up—sometimes after the fawn had been absent for as long as two or three days—and accepted her offspring with no apparent concern for any human scent.

The subject of supposedly orphaned fawns is an excellent place to introduce another common misunderstanding—that deer make good pets. Taking a fawn, genuinely orphaned or not, and raising it as a "pet deer" is a thoroughly bad idea that usually ends in great unhappiness for all concerned. There are several reasons for this. The first is that in most states it's illegal for unqualified people to keep wild mammals. Such laws exist for extremely valid reasons, which should be sufficient grounds for obeying them.

Second, what usually happens is that someone finds an "abandoned" fawn and raises it for a few months, even a year or two, and then the state wildlife agency hears about it. Conservation officers arrive and remove the deer over the outraged and tearful protestations of the family, and everyone loses. The unfortunate deer is no longer fit for release into the wild, the family is both sad and angry, and the wildlife agency people, no matter how carefully they handle the affair, are almost invariably accused of "Gestapo tactics."

Third, "pet" deer can be extremely dangerous. Although a great many people, especially those who are raising a so-called pet deer, simply can't believe such a lovely, gentle creature would ever harm them, the fact is that every year individuals are seriously injured, or even killed, by their gentle "pet."

Bucks are more dangerous than does, of course, because of their antlers, which can easily puncture a person. They're also apt to be especially rambunctious during the fall mating season. But even seemingly docile does can be unpredictable and turn on their keepers in an instant, slashing with sharp hooves that can do great damage. The bottom line is that wild animals, including deer, are just that—wild. Even though they may seem domesticated, they're still inherently wild, and no one should attempt to make pets of them!

A doe's milk is extremely rich—far richer than cow's milk—and fawns grow with astonishing rapidity. No doubt this is an evolutionary trait which ensures that fawns will be large enough by late autumn to survive the harshness of winter. Within a month, fawns have already tripled their birth weight and are also eating plant materials. After roughly two and a half months they can survive without their mother's milk, although they'll continue to nurse somewhat less frequently for another two or three months.

By late summer, fawns gradually lose their spots and begin to look much like smaller editions of the adults. Total weight by late autumn is largely a function of nutrition, but genetics certainly plays a role. In northern areas, fawns average seventy to eighty pounds, but they can grow much larger under ideal conditions. In rich farm country, where there's an abundance of high-quality feed all summer, fawns may run ninety to one hundred pounds after six months, and exceptional specimens may weigh more than 150 pounds.

The weights of adult whitetails, as well, are strongly affected by nutrition and genetics—but there's an added complication, known as Bergman's Rule. Simply stated, Bergman's Rule holds that members of a species grow larger as distance from the equator increases.

The theory behind this rule is that larger animals have less surface area in comparison to their body weight, thereby conserving energy in cold weather.

The scientific basis for this theory can be tested quite easily. Take a quart container, fill it with hot water of a known temperature, and put the cover on. Do the same with four half-pint containers, using water of identical temperature. Let the containers stand for an hour, and then check the temperature in the large container and one of the small ones.

Whether or not evolution was actually guided by this principle, northern deer are clearly much larger than their southern counterparts. A fully mature adult buck in a northern-tier state often weighs 250 pounds or more. In fact, the live weights of two enormous bucks from far northern Minnesota—one in 1926 and another in 1982—were estimated at an incredible 511 pounds! In contrast, 150 pounds might be more typical of a buck of the same age in, say, the Carolinas, while Key deer, a whitetail subspecies of southern Florida, tip the scales at well below one hundred pounds.

Despite the occasional super-whitetail, however, most deer are far smaller than people realize, especially in height. Although their slender legs and long, graceful necks may give the impression of height, most adult deer stand only three to three and a half feet high at the shoulder—which is a lot smaller than the impression they give. To put this in perspective, an adult whitetail with its head lowered can pass beneath the body of a full-grown moose.

Considering the enormous fascination that deer hold for humans, it's not surprising that a plethora of fallacies and myths have grown up around them. Some are obviously old, but others are of more recent vintage. The latter result from the steady migration of people off the land and into the cities and suburbs. Someone who has observed deer only in backyards, parks, and similar settings inevitably has less understanding of these animals' extraordinary capabilities than someone who has experienced them in their traditional wild milieu. Two examples demonstrate this point rather vividly.

During my days with the National Wildlife Federation, our headquarters received a letter from a very angry and upset lady who lived near a thirty- or forty-acre piece of land that had recently been donated to our organization. The previous fall, a forester from a highly reputable consulting firm had contacted me to discuss a small logging operation, which he would supervise. I had walked that land and felt that his proposal would improve both the wildlife habitat and the quality of the forest.

The letter writer, a recent émigré from suburbia, alleged that the beauty of the forest had been destroyed, and she feared that the deer would "break their delicate legs in the slash." Envisioning a logging job that had turned into a debacle, I phoned the forester, who assured me that the logger had done an excellent job.

In the end, the forester and I met with the woman, accompanied by two or three neutral parties, in an effort to resolve the dispute. What we found was one of the cleanest logging operations I've ever seen. The loggers had removed only the limited amount that the forester had described. All of the tops and slash had also been carefully lopped, so that nearly all of it was flat on the ground; most of what the woman termed "slash" actually consisted of a few slender saplings that had been bent over by the winter's snow.

This woman was far from stupid. In fact, I was quite impressed with the knowledge that she had acquired in a short time concerning deer in the locality. She had invested considerable time and effort tracing their trails to see when and where they moved through the area, and had learned much about their movements. Nevertheless, she had no frame of reference by which to judge either the logging job or the strength of a deer's legs.

Although she meant well, her fear that deer would "break their delicate legs" in almost nonexistent slash was wildly misplaced. Anyone who has seen much of deer in the wild knows better. I've witnessed deer bounding at high speed with perfect safety and apparent ease through logging slash so thick and deep that a person could barely clamber over and crawl through it. Likewise, I've seen whitetails racing up, down, and sideways at top speed along steep, rocky hillsides where a person could barely navigate.

In sharp contrast with the notion of deer as delicate creatures, I was presented with a graphic demonstration of their astonishing toughness and durability. While driving one day, I suddenly saw three does running over a rise to my left, and I had to brake sharply when they leaped across the road just in front of me. As I watched them bound down the bank to my left and across an open pasture to the woods beyond, a fourth doe suddenly flashed across the highway, following the route of the first three.

Near the base of the slope on the left stood the remnants of an old fence, a single strand of barbed wire still firmly attached near the top of a line of weathered cedar fenceposts. As the doe fled down the embankment, she turned her head to look back at me.

When she snapped her head forward again, the wire caught her under the chin in mid-leap. The result was quite astonishing: held by the chin, the doe swung like a pendulum so that she was upside down, head pointed back toward me.

The next instant, as the full force of her momentum took hold, the wire slid over her chin and she went rocketing through the air, bottomside up, to land on her back with a horrendous impact many feet away. The thought had no sooner flashed through my mind that this was one very dead deer when she whirled to her feet and galloped off into the woods, unscathed. So much for the fragility of the white-tailed deer!

Winter in northern portions of the whitetail's range is the time of maximum stress, when only the fittest may survive the combination of deep snow and bone-chilling cold. Despite the rigors of this season, however, deer are by no means defenseless against its perils. For starters, abundant autumn feed in good habitat allows the deer to put on a layer of dense, tallowy fat. The energy stored in this fat is a reserve that can be drawn on during the long, cold months, and may be crucial to survival in an especially severe winter.

Wildlife biologists realized this many years ago, and also learned that the very last of the fat reserves to be depleted are those in the marrow of the large bones. This led to a very quick, simple test to determine whether a winter-killed deer died of malnutrition and its associated effects or from some other cause. A biologist simply breaks open the femur (thigh bone), and inspects the marrow within. If the deer was well nourished when it died, the marrow will be white and fatty, almost like suet. If the deer was moderately malnourished, the marrow will have a pinkish tinge and will be somewhat less solid and fatty. But if the deer was badly malnourished, the marrow, lacking any fat, will be thin and red—almost the color and consistency of red jello. In fact, the marrow of a deer in the final stages of malnutrition is so thin that it can be spread on a printed page and the print read right through it!

A deer's second defense against bitter cold is its winter coat. Twice each year—once in the spring and again in the fall—whitetails molt, much as birds do. That is, they gradually shed one coat and replace it with a different one. During the spring molt, in particular, the winter coat comes off unevenly, while the summer coat is replacing it.

The result is that whitetails in late spring or early summer often have a very mangy, moth-eaten look. This passes quickly, though, and the deer are soon sleek and resplendent in a glossy, reddish summer coat. This consists of only a single layer of fine, solid hairs that shows its owner off to great advantage.

When autumn arrives, the process is reversed: off comes the summer coat, replaced by winter garb. This winter coat is a marvel of efficient heat retention. A thick, soft undercoat holds in body heat, and what heat escapes that layer is retained by the outer coat. The latter consists of longer hairs, each one of which is hollow—an air trap that provides outstanding insulation. In addition, the winter coat is a grayish brown, much more suited to the somber color of the fall and winter woods than the red summer coat.

The third winter defensive strategy is to move into winter quarters. These are popularly called "deer yards," which is something of an unfortunate misnomer, since it gives people the impression of deer crowded into a little area much like a sylvan barnyard.

This misunderstanding manifests itself in a variety of ways. As a classic example, Walter D. Edmonds, in his 1936 novel, *Drums Along the Mohawk,* an otherwise well-researched book, described a scene in which the settlers went to a deer yard and, when the deer stopped against the far wall, used them for target practice. This description belies reality, which is entirely different.

Biologists much prefer the term "deer wintering area" to "deer yard," because it's far more descriptive of actual conditions. Even in deep snow, a deer wintering area has packed paths or trails scattered through it, and it's often possible to wander around in one of these areas for hours and catch little more than a fleeting glimpse of a deer. There are such things as very small wintering areas, where it's likely one will get a better look at the inhabitants, but a wintering area that small will normally contain only a handful of deer. In any event, so-called deer yards don't have "walls" of snow, or anything else.

Deer have very specific requirements for wintering areas, and if the winter habitat is destroyed—by clearcutting or a housing development, for example—the deer can't simply go and find a replacement. The first requirement in areas of severe winters and heavy snowfall is low-elevation, mature softwood growth, usually at least forty years old. The thick, interlocking canopies of mature cedar, spruce, balsam, pine, or hemlock trap much of the snow as it descends; subsequently some of this snow evaporates, and what filters down to the ground is so fine that snow depth is much reduced. This, of course, greatly facilitates travel by the deer.

In addition, the dense evergreens break the wind and cut wind chill a great deal, and they reduce radiational cooling at night. There may also be an added thermal benefit from mature softwood cover, because the dark evergreens absorb heat from the sun and make a small but crucial difference in the temperature below.

A second attribute of a good wintering area is a south- or southwest-facing slope so that the low-lying winter sun in afternoon strikes more directly, and hence provides more heat. Deer also manage if the terrain is flat, but won't select wintering areas on north- and northeast-facing slopes.

The whitetail's final defense against winter is a reduced metabolic rate. This helps them conserve energy when food is scarce and the cold is severe. Even with all these adaptations, however, winter is a dangerous and often deadly time for whitetails. In northern climates, even under good conditions, they lose a substantial amount of weight, and when that loss approaches one-third of their body weight, death from outright starvation and causes related to severe malnutrition begin to set in.

A prime wintering area contains patches of mature softwoods interspersed with smaller openings that provide younger trees for browse—ideally, hardwoods or white cedar. However, because of their reduced metabolic rate, whitetails can endure several weeks in winter with little or no food, provided they aren't disturbed. As soon as deer become nervous and agitated, their metabolic rate rises; if the deer are forced to take flight, energy consumption becomes even greater. That's why winter recreationists such as cross-country skiers, snowshoers, and snowmobilers should avoid deer wintering areas.

Despite all these adaptations, a harsh winter will still take a substantial toll of whitetails. Some deer die even in a mild winter, but many fawns, as well as some older deer, succumb during a severe one. In fact, life in the wild at any season is extremely hazardous for deer. Ask a room full of people how long they think an *average* deer lives in the wild, and many will guess fifteen years, twenty, or even more. They're grossly overestimating. Although a small percentage of deer live into their teens, the vast majority die much sooner, and the average life span of a whitetail is only about two years!

No account of the hazards of winter would be complete without mention of the damage caused by roaming domestic dogs. When confronted by the fact that their small dogs are "running deer" (the colloquial phrase for pursuing deer), many owners bridle and say, "Why, Fido and Fifi wouldn't hurt a fly—and besides, they're far too small to damage an animal as big as a deer."

This is a huge misunderstanding. True, someone's little terrier, spaniel, or beagle can't attack a deer and bring it down by brute strength, but it can easily run a deer to exhaustion in the winter, especially if the snow is deep. When a deer finally drops from fatigue, even small dogs may begin to gnaw on it while it's still alive—and even if they break away from the chase and head for home, an exhausted deer is often so stressed that it won't recover.

Roaming dogs taught our children at an early age a harsh lesson in the damage that dogs can inflict on deer, as well as in the hard realities of the natural world. In order to take the school bus, the children had to walk a mile down our narrow, wooded dirt road. One day, halfway down the hill, they came on a deer lying in the road, completely exhausted by pursuing dogs. It was an appalling spectacle; the deer was still alive, and the dogs were already beginning to feed on it. The children ran back to the house to fetch my father (I was away at the time), who drove the dogs away, shot the deer to put it out of its misery, and called the game warden.

One common myth about deer in winter is that starving deer can be saved by giving them hay or grain. Deer are ruminants—that is, they chew their cuds and have a stomach with four segments—and ruminants depend on microorganisms to help them digest their food.

Winter microorganisms in a whitetail's stomach are different from those which help it digest summer food. In summer the microorganisms are varieties that can digest a wide array of herbaceous plants. In winter, however, the whitetail's normal food supply consists of browse—the tips of small twigs and branches—and the microorganisms in its digestive system shift accordingly to types equipped to handle this woody fare. That means that a malnourished deer in the winter simply can't digest these new foods presented to it by well-meaning humans and will literally starve to death with a full stomach.

Winter feeding of deer *can* be successful, but only if it's started early enough in the fall. Whether or not it's a good thing, however, is highly debatable. An increasing number of people are now buying "deer pellets" at their local feed store and supplying them to deer in their backyards. While people understandably enjoy seeing these beautiful animals close to their homes, and feel good about saving some of them in a severe winter, artificial feeding of deer has two very undesirable results.

First, artificial feeding tends to exacerbate what's often an overpopulation of deer to begin with. As overpopulation grows worse, more and more deer have to be fed ever larger quantities of expensive food to keep them alive. Eventually, things reach the point at which deer become an intolerable nuisance, even though the artificial feeding may be able to sustain them.

A second problem caused by artificial feeding is far less tangible: it robs deer of their wildness. One of the most splendid native traits of the white-tailed deer is its quintessential wariness. To many of us, this deep, instinctive caution is the quality that makes deer the magnificent animals that they are. But deer are highly adaptable; they soon learn the pleasures of handouts, and rapidly adjust to them. In the process, they also adjust to human proximity and become semi-tame—a sort of oversized yard pet. This is degrading both to the deer and to us. It robs them of a vital spark at the very core of their nature, and it makes us a party to this process simply because we want the gratification that comes with seeing semi-tame deer around our houses. Although some argue that feeding deer is no different from feeding birds, there's one essential difference: birds don't lose their wildness by being fed, but deer inevitably do.

Deer aren't very vocal most of the time, but they're capable of making several different sounds. The one most commonly heard, perhaps because it's the only one that carries for any distance, is a high, whistling snort. This sound is often referred to as "blowing." Blowing or snorting represents an alarm call, a warning to other deer. Mostly it's used when a deer suspects something is wrong, but isn't quite sure. As soon as the certainty of danger sets in, the deer usually bolts, although it may run a short distance, pause, and snort over and over for a number of minutes.

Anyone unfamiliar with this sound can readily approximate it. First jut the lower jaw forward to create a narrow opening in the mouth. Then expel your breath with explosive force against the back of the upper teeth and use your diaphragm to maximize the amount and velocity of breath being expelled. Every effort should also be made to pitch this sound as high as possible. The result, at least with a little practice, should give a fair idea of what a blowing whitetail sounds like.

The bleat of a hungry fawn has already been described, and I've also heard a fawn with its mother make a little squeaking sound, almost like the sound of creaking saddle leather. Does bleat also, and can on occasion make a louder sound. One night, just as my wife was about to turn into our driveway, a doe leaped off our lawn, landed on the hood of the car, rolled off, and galloped away unhurt—but not before she expressed her dismay with a loud *BAAAAA!*

Bucks, in addition to the snorts that all whitetails use, also grunt in a guttural fashion, not unlike a pig, although the sound doesn't carry as far. In particular, a buck hot on the scent of a doe in heat will often utter a series of what's known as "tending grunts."

The past four hundred years have given the white-tailed deer a strange rollercoaster ride. From great abundance in a land of primeval forests, hunted by wolves and cougars with some human predation thrown in, they soon dwindled to a tiny fraction of their former numbers in a land changed beyond recognition. Gone were most of the forests, gone were the wolves and cougars, replaced by a flood of human predators.

Then the tide turned. Forests returned, and the human predators that had once decimated the deer now helped to restore them. Up, up went the deer again, to superabundance undreamed of a century before. No longer merely numerous, they now present a problem of excessive population in many areas. Beyond sheer numbers, they've become interwoven into the fabric of our daily lives in ways ranging from things as prosaic as dollars and cents to emotional and aesthetic meanings too deep to fathom. As the most visible and important proof of restored wildlife, and as a symbol of the wildness which so many of us treasure, the white-tailed deer surely deserves the title of the Comeback Kid.

23

The Great Stripper: The Moose

MYTHS

- Moose are slow-moving and ponderous.
- Moose are quite tame and safe to pet.
- Moose drive out deer.

HUGE, BULKY, AND NEARLY BLACK, MOOSE *(ALCES ALCES)* ARE SPECTACULAR CREATURES. With the possible exception of a bear, no other North American mammal can draw a crowd of curious and excited spectators as quickly as a moose. Let a moose be sighted, and soon cars line both sides of the road, disgorging crowds of eager people with cameras and binoculars. If ever an animal deserved the title—now something of a cliché—"charismatic megafauna," it's surely this giant beast.

Moose are believed to have arrived in North America via the Siberian land bridge about ten thousand years ago, during the last ice age, and they've been here ever since. They're very much a northern animal: it's estimated that over 700,000 moose—about 80 percent of our continent's moose population—reside in Canada, with many of the remainder found in Alaska. In the lower forty-eight states, moose are mostly restricted to the northern Rocky Mountain states, the very northern portions of the Great Lakes states, and northern New England and New York.

Although moose numbers have historically remained fairly stable—and perhaps even increased—throughout much of their range, the situation has been vastly different in New England and New York. Archaeological evidence indicates that moose were the dominant large herbivore in that region when the first European settlers arrived. As far as those settlers were concerned,

White-tailed deer *(left)*; moose

they had landed in fearsome wilderness that had to be tamed and made to look as much as possible like the England from which they had departed to seek religious freedom (which they promptly denied everyone else) and the opportunity for greater prosperity.

But the business of wilderness-taming was hard, hungry work, and the settlers dined well on a variety of wild game—such animals as wild turkey, white-tailed deer, and moose. Just as moose were highly prized by the Native Americans, the settlers quickly learned to appreciate the huge animals for their delicious meat and valuable hides. Lazy these settlers were not, and they pitched into the task of "civilizing" their environment and putting game on the table with astounding energy. Thus it was that with ax and saw, fire and gun, the settlers inexorably eliminated most of both moose habitat and moose from the Northeast.

Despite the passage of laws from the late 1800s to the 1930s, giving moose complete protection, and the farm abandonment and forest regeneration that fueled the amazing comeback of the white-tailed deer, the recovery of moose in the Northeast was very slow. In fact, it really wasn't until well after World War II that moose numbers in the region began to increase appreciably.

The trigger for the moose's upward climb was a radical change in logging practices. Even the wilder parts of the Northeast weren't true wilderness by the late nineteenth and early twentieth centuries, for they had already been logged several times. However, timber-harvesting methods concentrated on bringing the logs and pulpwood to the transportation system; that is, horses dragged timber from cutting areas to rivers or roads for transportation to the mill. This was a time-consuming system that tended to limit the speed with which areas could be cut.

All that began to change after the Second World War; now the strategy shifted to bringing the transportation to the logs. Huge bulldozers began to push wide roads farther and farther back into what had essentially been rather wild and inaccessible country. Logging became increasingly mechanized, for the bulldozers were followed by log skidders, tree harvesters, and trucks.

As a result, huge areas covering many hundreds—even thousands—of acres were clearcut. In a very few years those clearcuts began to grow back, mostly to the young hardwood trees which are prime moose fodder. In short, the new logging practices created vast "moose pastures" that the moose were quick to take advantage of. Moose numbers began a steady upward climb, first across northern Maine, next across northern New Hampshire, and finally across northern Vermont and New York. Then the great moose expansion began to spread southward, gradually encompassing most of the territory in the three northern New England states.

Of course, logging wasn't the sole reason for the resurgence of moose in this area. Rapidly increasing numbers of beavers (see chapter 1) helped, and the reduced number of white-tailed deer in areas of huge clearcuts may have played a role, as well. This will be explained a bit later.

Although growing numbers of moose have delighted many people, not everyone is thrilled. Undeniably, large numbers of moose can cause significant problems, owing partly to serious misunderstandings about the real nature of moose. As perhaps befits their great stature and bulk, moose tend to go right through obstacles, rather than ducking under or leaping over them in the fashion of their smaller brethren, the whitetails. This causes major headaches for maple syrup producers who find their sap pipelines knocked down by moose passing through their sugar woods; likewise, farmers quickly become weary of repeatedly chasing after cattle that have wandered from their pastures through gaping holes in fences torn by wandering moose.

As aggravating as these problems can be, they aren't the most serious ones. As the numbers of moose increase, so do the collisions between automobiles and moose. Unlike collisions with deer, which can cause expensive damage to a car but mostly don't seriously harm the occupants, collisions with moose are extremely dangerous. Because of its long legs, the body of a moose is higher than the hoods of most passenger cars. The result is that when car strikes moose, the body of the moose usually smashes into the windshield or comes down on the roof, flattening it. Often driver and passenger are seriously injured, and a number of people have been killed in these crashes.

To some degree, such collisions can be attributed to the highly erroneous idea that moose are ponderous and slow-moving. True, their extremely long, almost stiltlike legs and great bulk give them an awkward, ungainly look—but no one should be fooled by this highly deceptive appearance. Moose can sustain a ground-devouring trot of five to ten miles per hour for hours on end, and at full throttle they're capable of going at least thirty-five miles per hour—nearly as fast as a whitetail's top speed! Contrary to the expectations of those who think moose are slow and lumbering, they can intersect with a speeding automobile with astonishing rapidity.

My wife and I experienced a graphic demonstration of this fact four or five years ago. We were driving on New Hampshire's Kancamagus Highway through a part of the White Mountain National Forest. It was broad daylight, and we were cruising at the fifty-mile-per-hour speed limit along the downslope on the Maine side of the pass. There was a long, straight stretch at that point, the road was fairly wide, and beyond each shoulder was a ditch, with a cleared strip between it and the woods.

Suddenly I caught a movement out of the corner of my left eye. Something about the movement suggested that it was bigger than a bird, and the thought flashed through my mind that it might be a hiker emerging from the woods. All that consumed only a split second, and when I glanced in the rearview mirror, there was a big cow moose, her hindquarters sprawled across the centerline and her head pointed back to my left. I yelled, "Moose!" to my wife, and just as she turned to look, we both saw the animal leap to its feet and speed back into the woods in the direction from which it had just come.

It wasn't difficult to reconstruct what had happened. The moose must have shot forth from the woods with her throttle wide open. She crossed open strip, ditch, shoulder, and part of the highway in a flash, nearly hitting the back of our car. At that point she must have slammed on her brakes and skidded on the asphalt surface. Her hindquarters went down, slid, and pivoted her 180 degrees. It was a remarkable and frightening display of this "slow" animal's speed!

If this story seems unlikely, just consider the mathematics of it. Sixty miles per hour equals eighty-eight feet per second. A car traveling at fifty miles per hour moves at seventy-three feet per second, and a moose in a hurry covers ground at roughly forty-four feet per second. The flicker of movement that I caught out of the corner of my eye meant that the moose had started to emerge from the woods slightly ahead of our car. The moose would have covered the distance to the edge of the travel lane, approximately fifteen feet, in one-third of a second; meanwhile, our car would have traveled about twenty-four feet, so that the rear of the car would barely have passed the moose as it reached the pavement. Then the moose slammed on its brakes and skidded as it hit the pavement.

Just as I was printing the very last chapter in this manuscript, the phone rang. It was our daughter Suzy, who asked in a very shaky voice, "Can you come and get me? I just hit a moose!" A yearling bull moose had come charging out of the woods, directly into her path. Thanks to her quick thinking, she'd avoided injury; thoroughly indoctrinated with the knowledge that she could be killed in a head-on collision with a moose, she had slued her car sideways so that she struck the moose only a glancing side blow. Unfortunately, the moose suffered a badly broken front leg and had to be dispatched by the game warden, but there's a long waiting list for moose meat, so the meat, prime at that age, was salvaged.

Suzy had heard time and again about the speed of a running moose and knew, in an intellectual sense, that they could emerge from the woods at a frightening clip. The reality, however, was a revelation that she summed up by saying, "I still can't believe how fast they move!"

May is the worst month for moose-and-car collisions, at least in the more southerly portion of the moose's range. Here there are plenty of paved highways where salt is used in winter to melt snow and ice. This salt ends up in wet areas beside the road, creating artificial salt licks. Such sites are easily identified: they resemble a wet barnyard just traversed by a herd of cattle, the water muddy and the ground a soggy mass of great hoofprints.

Although moose come to these salty places all through the warm months, they use them far more heavily in May. Why? Because the aquatic plants that supply moose with sodium and other nutrients haven't yet grown enough that moose can feed on them. With moose heavily concentrated during May along paved highways, and cars zipping along these roads at high speed, it's inevitable that collisions disastrous to moose, car, and occupants will occur.

Moose are especially hard to see at night because of their very dark color; often the first thing a motorist sees are the long gray legs, which show up a little better than the body. Moose love the accumulation of winter road salt that collects in wet areas beside highways, and this makes nighttime travel doubly hazardous along these "salt licks."

Although it's advisable to slow down when driving through moose country, especially at night, that's no guarantee against a collision with a moose. I've personally known of several people who slowed down substantially because they were afraid of hitting a moose—and hit one anyway!

People often suggest mooseproof fencing along highways to prevent accidents. Proper fencing does, in fact, work very well, but it's prohibitively expensive for all but very short stretches of highway. For example, in 1987, Alaska fenced a ten-mile section of high-speed highway out of Anchorage, where there were many collisions with moose each year. The strategy worked, but it cost $1.25 million just for that ten miles, and the cost today would probably be at least 50 percent higher. Clearly, fencing isn't an option except in the most unusual circumstances.

A third serious problem is due almost entirely to human idiocy, which sometimes knows no bounds. At least partially because of the influence of Bullwinkle, the cartoon character, far too many people regard moose as oversized woodland bovines, placid, benign, and cuddly. As a result, they constantly put themselves in danger by trying to approach moose too closely—or even pet or hug them!

My files contain numerous accounts of people who have been killed or seriously injured by moose, as well as other examples of individuals who took horrible—and inexcusably stupid—chances and were fortunate enough to walk away unscathed. Evidently, fortune sometimes favors fools.

One of the most egregious examples concerns a woman attempting to pet a "tame" moose that had been hanging around, attracting crowds of spectators. A newspaper photo showed her reaching up to pat the cow moose: the moose, registering its extreme displeasure, had her nose pointed skyward and her ears laid back nearly flat, much in the manner of an angry house cat. That was bad enough, but the caption informed us that the woman had attempted this earlier and been kicked by the moose! This is enough to make one wonder if humans actually do sit atop the scale of animal intelligence.

Even those who should know better are sometimes a party to such dangerous foolishness. For instance, just last year the magazine of a national conservation organization featured a two-page color photo of a man in Quebec hugging a moose. Several letters to the editor pointed out that the photo would only encourage others to engage in such dangerous behavior and urged that some kind of retraction or warning be printed. There was no reply, which surely tells us something about their sense of editorial responsibility.

Moose are large, powerful, wild, and unpredictable, and hence potentially very dangerous. They should *never* be approached closely except by a trained and qualified person, such as a biologist or conservation officer. Unpredictability is the key word here. Yes, there are times when moose act exceedingly tame, and a person is able to walk up and touch one with impunity. On other occasions these seemingly tame moose can lash out unexpectedly, with decidedly detrimental results to the human involved. The problem is that there's no way to tell in advance which way a moose will react.

Such apparent tameness may sometimes result from disease, but in other cases there's no known reason. In any event, this sort of docility represents aberrant behavior in moose, which generally shy away from close human contact. It is true, however, that moose may become accustomed to humans in certain situations and ignore them unless they approach too closely. Ponds frequented in summer by both moose and people in boats are a good example. In that sort of setting, a moose may allow a boat and its occupants to come within a hundred feet or so without any evidence of concern. In nearby woods, however, the same moose will flee from humans at a far greater distance.

Even in situations where moose are acclimated to human proximity, too close an approach can be dangerous. The owner of a fishing camp in northern Maine, where boats routinely come reasonably near feeding moose without difficulty, told me about one unfortunate incident in which two or three fishermen in a boat decided to get considerably closer to a feeding moose than they should have, for the purpose of obtaining some good photos. Without warning, the moose charged them and overturned their boat. A great thrashing about of moose and men ensued in the shallow water. Luckily the men escaped serious harm, but they emerged from the water with a much healthier respect for moose!

There are also cases where moose are not at all tolerant of human presence. Sometimes the reason for aggressive moose behavior is unknown (unpredictability strikes again), but cows with calves can be very protective of their young, and hence highly dangerous; it's especially perilous to come between a moose and her calf.

Bulls in rut, stoked by rampant testosterone levels, are also very dangerous—as I well know from personal experience. My most terrifying encounter with a wild animal involved a bull moose, and it happened in this fashion.

I was staying at the aforementioned fishing camp during late September, near the height of the moose's mating season. One day it was raining and too windy to fish. Feeling overfed and underexercised, I set out for a brisk walk of several miles on the road, owned by a paper company, that provides access to the camps. About halfway back to camp there was a large clearcut on the right of the road and woods on the left. On the left, between the woods and the road, stood a dense growth of tall brush whose thick foliage totally obscured a view of anything within.

I was walking rapidly, when suddenly a soft but deep grunt sounded almost in my left ear. Instantly I realized what it was and thought, *Oh God, not that!* Before I could even react, a second grunt, almost piglike, followed, and then the bushes next to me began to move. I'll readily confess that I panicked. I should have run down the road, cut into the woods on the left, and climbed a tree. Instead, I did what instinct told me to do and fled directly away from the sound, straight into the big clearcut on my right. Bad move!

After running about a hundred yards, I glanced back. To my horror, there was a large bull moose with wide antlers coming after me! In that instant, I was certain that I was a dead man and was terrified beyond words. There was nothing large enough to climb in the clearcut, but there were a few little clumps of brush here and there. I bobbed and weaved though three or four of these, all the while circling back toward the road. Finally I threw another look over my shoulder and, to my inexpressible joy and relief, there was no moose in sight.

I emerged onto the road farther away from camp and had to walk back past the spot where the big bull had been hidden. It was all I could do to muster enough courage to pass the spot, and I did so only with great trepidation, the utmost caution, and the firm conviction that if further flight became necessary, it would be into the woods and up the nearest climbable tree.

This adventure had what, in retrospect, was an amusing sequel, although it certainly didn't seem it at the time. I was nearly back to camp and was just beginning to relax when, with a thunder of wings, two ruffed grouse exploded out of the roadside underbrush no more than three feet from me. I leaped two feet in the air and came down with my heart racing at a horrendous rate!

It wasn't until weeks later, after the terror of this incident had diminished, that I realized what the bull moose was actually doing. A moose can easily outrun a human, and that bull could have caught me and stamped me into the ground in fifty yards if he had chosen to do so. He resented my presence in his territory during the rut and was chasing me, much as he might have a vanquished rival bull (I was certainly vanquished, but not a rival!) to ensure that I left the area. As soon as honor—and testosterone—were satisfied, he broke off the chase.

It has always struck me as ironic in the extreme that someone like me, who has constantly warned others about the dangers of approaching wild animals too closely—particularly large animals like the moose—should have been chased by one. The fact that I had no way of knowing about the bull's presence also underscores the fact that nature is unpredictable and often dangerous; consequently, we should banish from our heads any warm, fuzzy thoughts about "benign nature."

Moose are often called homely and ungainly, but that does a grave injustice to one of nature's finest exhibits of adaptation, and they might better be termed imposing and majestic. As an example of adaptation, consider the food requirements of the moose. A large adult moose needs fifty to sixty pounds of vegetation daily in order to feed its huge frame. If a moose had to consume that quantity of browse by nibbling at the tips of twigs in the fashion of its whitetail relatives, it would have time for little else.

The animal's name comes from the Algonquian *moosu,* which has been variously translated as "twig eater" and "he who strips off (leaves and bark)." My preference is for the latter because it's such an apt description. The big, rubbery upper lip, which is partially responsible for the moose's "homely" appearance, is in fact a most useful adaptation: the moose simply wraps its lip around a twig and, with a deft motion, strips away leaves, bark, and buds.

This improved browsing method enables moose to eat their fill with exemplary speed and efficiency.

Likewise, the long legs that give the moose its "ungainly" appearance are admirably suited to life in the harsh climate that makes up the moose's habitat. Those legs are not only long but also extremely powerful, and they easily drive their owner through deep snow that would mire a deer.

As huge as the moose seems to us, it's merely the runner-up in the deer family's size sweepstakes. An extinct relative from the last ice age, the somewhat misnamed Irish elk *(Megaloceros giganteus)* was the largest deer that ever lived. Its enormous antlers—up to twelve feet wide (some sources say up to fourteen feet)—were somewhat palmate (flattened, with fingerlike projections), although not as broadly so as those of a moose.

The species received its common name because the best-preserved specimens have been found in peat bogs and lake silt in Ireland. This giant deer became extinct after the end of the last ice age, and the latest known specimen of this truly remarkable beast dates back about eleven thousand years.

Extinct giants notwithstanding, our present-day moose is enormously impressive in its own right. How large is a moose? Of the seven subspecies of moose worldwide (the European subspecies is called elk), the largest is believed to be the Alaskan moose. There are various ways to measure the size of a moose. Weight is one of them, and a huge bull moose can weigh 1,600 pounds. Then there is height. A large moose can stand at least seven and a half feet tall at the shoulders, and the tips of the antlers on a big bull may tower a full ten feet above the ground. All of this is accompanied by strength commensurate with the immense height and bulk of its owner.

Antler development is equally impressive as a measure of size. Moose antlers can stretch as wide as six and a half feet and weigh well over fifty pounds while the moose is wearing them (shed antlers dry out and weigh substantially less). One of the mysteries of nature is how a moose can run full tilt through dense forest without constantly banging those great antlers against trees and other obstructions. Evidently a bull moose has an astonishing sense of spatial relations wired into its genetic makeup to give it such uncanny ability.

As with the antlers of their deer relatives, moose shed those great appendages every winter and grow another set, starting in the spring. Moose antlers, like those of deer, are quite tender during the growing period, covered with velvet and full of blood vessels and nerves.

A yearling bull will usually produce a set of spikes, just as many yearling whitetail bucks do. The following year, the two-year-old bull will grow small palms. Thereafter—health and other considerations being equal—his antler size will increase year by year until he reaches his prime. Then antler size levels off and remains more or less constant for a few more years. If the bull survives past his prime, antler size will begin to diminish as age takes its toll.

In general, ages five to ten years are considered the prime of life for a bull, though a few exceptional individuals may prolong that for another two or three years. These exceptions are likely the result of a combination of good nutrition, excellent health, lack of serious injuries, and genetics. Moose in general are substantially longer-lived than white-tailed deer. They often live well into their teens, and a few may even reach the age of twenty, so an occasional bull a dozen or so years old may still manage to be the biggest kid on the block, so to speak.

With the diminishing sunlight of late September, testosterone levels in the bulls rise accordingly, and their huge antlers come into their own. Now the mature bulls are just spoiling for a fight in order to show their dominance and claim a small harem of cows. Doubly unpredictable and dangerous during the rut, feisty bulls have been known to attack strange objects; there have even been reports of a bull moose charging a railroad locomotive!

One of the employees at the fishing camp where I was chased by the bull moose told me about a sight he had witnessed. A big bull that was causing a great commotion across the pond attracted his attention, so he hopped into a boat and went over to observe the show—from a discreet distance, I might add, for he wasn't stupid. He reported that the moose, evidently frustrated by the absence of any rival bull with which to quarrel, took out his anger on the surrounding vegetation. It was an awesome display of power. The bull thrashed his antlers this way and that, on the ground and above it, stamping around and grunting all the while. Then he began to dig his antlers into the ground and uproot clumps of alders, tossing them into the air over his head. All in all, it was a bravura performance.

According to those who have witnessed it, a fight between two big bulls is even more spectacular. Smashing their antlers together, the two leviathans shove back and forth in a test of strength, tearing up the ground for yards around. If they're evenly matched, this battle for dominance can go on for quite a few minutes until one of the bulls is convinced that the other is bigger and stronger. Then the vanquished bull will break away, turn, and flee. The victor is apt to pursue him hotly for a short distance, often goring his erstwhile foe in the process, but breaks off the chase as soon as he's sure his rival is well and truly beaten.

Although bull moose are polygamous, they breed far fewer females than does a dominant whitetail buck. Perhaps this has something to do with the fact that moose, being so much larger than deer, have a lower population density, and thus would have to travel much greater distances to find so many females.

One means by which bulls locate cows during the rut is by construction of pits, often called "wallows." The bull first scrapes out a shallow hollow in the earth, then urinates in it and spreads scent from glands on his legs. Finally he rolls about in the malodorous affair, liberally coating himself with its scent. These wallows attract cows by their pungent odor; a bull will have several of these moose equivalents of a singles bar scattered about his territory, and will visit them frequently in search of females in heat. The wallows' aroma that the bull attaches to himself may also attract cows and put them in an amorous mood.

Not long after the rut is over, moose begin to seek their winter quarters. Both winter quarters and wintering habits of moose are very different from those of white-tailed deer. Whereas deer tend to winter in large areas containing many deer, moose spend the cold months in small groups in equally small pockets of good winter habitat. Two or three moose—rarely as many as six or seven—seek small groves of softwood trees located near plenty of good hardwood browse. Also, in latitudes where the ranges of moose and whitetails overlap, moose generally winter at higher elevations than the deer.

Strange as it may seem, moose actually need the shade of the coniferous trees in late winter to prevent them from overheating. No doubt their very dark color acts as a solar collector and contributes to overheating when the sun's rays grow more direct with the approach of spring.

Most cow moose are bred around the first week of October. With a gestation period of a little under eight months, that means most calves are born from mid-May to the first of June. About 20 percent of the yearling cows breed; these almost invariably have a single calf. Likewise, most of the two-year-olds bear single calves. Cows from ages four to nine are considered to be the prime breeders, and about 40 percent of those cows have twins. Triplets occur, but are rare.

The calves weigh twenty-five to thirty-five pounds at birth. Although their long legs are wobbly at first, they can follow their mother around after only three or four days. Like whitetails, moose have very rich milk, so the calves gain weight at a phenomenal rate. They put on a pound a day for the first month and two pounds a day for the next three or four months. As a result, these huge "babies" often weigh three hundred pounds or more by December—far more than most mature whitetails. Although the calves can survive on their own after two months if something happens to the cow, they normally continue to nurse until October. By the following October, the year-old moose will weigh a remarkable four hundred to six hundred pounds!

Summer food for moose is quite different from their winter fare. Because of a winter diet consisting almost exclusively of browse, moose emerge from the long winter months deficient in sodium and certain other elements. To remedy this lack, succulent aquatic plants now become an important part of their diet. These plants are rich in sodium, iron, and other nutrients, and moose begin to feed on them eagerly as soon as they become available.

Good moose habitat contains plenty of wetlands—shallow, boggy lakes, beaver ponds, and similar areas. These serve a dual function: first, they provide a rich source of aquatic plants for food; second, they keep the moose cool in hot weather and offer some relief from the hordes of biting insects that torment them at that time of year.

Moose are powerful swimmers, so at home in the water that they might almost be called semiaquatic during the summer months. They have hollow hairs that provide considerable buoyancy, so, despite its great bulk, a moose has little difficulty staying afloat. Most of a moose's aquatic time, however, is spent wading in relatively shallow water, where it can submerge head and neck in order to bring up favored plants to be munched above the surface.

Thanks to this trait, my wife and I had a memorable experience with a moose. We were canoeing the shoreline of a pond, looking for moose while the September dusk was starting to settle. Suddenly we began to hear what sounded like a waterfall, although we knew there were none nearby. Moreover, the sound was sporadic, rather than the steady noise made by falls. Suspicion dawned on us, and we paddled hurriedly toward the source of the sound, along the far shore some distance behind us.

Sure enough, there was a huge bull moose, one of the largest we've ever seen. Although he was deeply engrossed in feeding and paid no attention to us as we approached, we maintained the properly respectful distance befitting such a forest monarch. Each time he raised his head from beneath the water, his mouth stuffed with plants, water cascaded off the wide palms of his majestic antlers, temporarily imitating the sound of a miniature waterfall.

One of the most misunderstood aspects of moose biology is their interaction with deer. Over and over, the refrain is heard in some circles, "Moose drive out the deer." When people correctly observe that there are more moose and fewer deer in a particular area, they frequently make the erroneous assumption that this is a clear case of cause and effect. It isn't.

There is no physical conflict between moose and whitetails; the two species go their separate ways and don't bother each other. Any effect that moose have on deer, then, real or imagined, must be related to food supplies, and there are at least two reasons why this is rarely a problem.

First, although moose and deer feed on the same species of trees and shrubs, moose mainly browse at a considerably higher level than deer. Second, summer browse isn't usually a problem; it's browse in the wintering areas that's critical. As already noted, moose winter in small groups at elevations higher than most deer wintering areas, so there's seldom competition between the two species for winter food. There have been a few documented cases of competition for winter food between moose and whitetails, but these are very uncommon.

Then why do deer populations often decline while moose numbers rise in the same area? The answer lies in habitat change—but change caused by humans rather than moose. As previously described, wholesale clearcutting has benefited moose greatly by creating a sea of browse, often called "moose pastures." Frequently these huge clearcuts destroy deer wintering areas by removing the mature softwood stands so critical to whitetail survival in severe winters. Without the shelter of these softwoods, deer flounder in the deep snow, and many perish, while the powerful, long-legged moose continue to move freely. As a result of this major habitat change, moose thrive and deer decline.

Far from being driven out by moose, white-tailed deer can actually have a harmful effect on moose. Enter the villain, a diminutive roundworm with the outsized name of *Parelaphostrongylus tenuis,* better known as the brainworm. The interrelationship between deer, moose, and brainworm is complex, yet sufficiently fascinating to be well worth exploring.

Brainworms, threadlike parasites less than four inches long, are found in the brain covering *(meninges)* of the majority of deer. However, deer seem unaffected by their presence. Eggs of this parasite travel in the bloodstream to the lungs, where they develop into larvae. The larvae then move up to the throat, where they're swallowed and eventually pass out of the deer in the feces.

This would be the end of the cycle—and the brainworm—if it weren't for snails and slugs. These pick up the larvae from the deer feces, and the larvae then develop further inside their new host. Eventually the snail or slug, brainworm larvae and all, may be accidentally ingested by a browsing moose.

Once inside the moose, the larvae migrate to the spinal cord. There they mature into adult brainworms and travel to the brain. Then the parasites attack the brain itself, rather than living in the meninges, as they do in deer. The result is a condition known as "moose sickness," which is nearly always fatal.

Moose afflicted by this condition display a variety of symptoms. They may act very tame and lethargic. (No doubt this is the source of many of the "tame" moose that people foolishly try to pet.) Later symptoms may include such things as lack of coordination, walking in circles, and even blindness. Finally the moose becomes paralyzed and dies.

Because deer are the ultimate source of brainworm infection in moose, a high population of deer can inhibit moose numbers. When deer populations decline as the result of large-scale clearcutting, moose benefit not only from vast areas of browse, but from fewer deer, as well. Thus if there is any interaction between the two species, it's the deer that are more likely to drive out the moose, no matter how indirectly, rather than vice versa.

Brainworm isn't the only affliction suffered by moose. There is also the winter tick. This parasite sometimes infests moose in such quantities that *thirty thousand* or more of them can be found on a single animal! As moose try to dislodge these pests, they scratch with their hind hooves and repeatedly rub against trees. In the process, they may remove substantial portions of their outer coat; especially in calves, this added exposure to winter cold may prove fatal.

Moose are less vocal than deer throughout most of the year, primarily because they lack the whitetail's high, whistling snort of alarm. The bulk of their noisemaking comes during the rut, when the bulls utter their deep grunts. At that time, cows facilitate the process of finding a mate by uttering high-pitched, moaning cries and what's sometimes described as a bellow.

Calves aren't entirely silent, either. One late September evening, my wife and I were in a canoe watching a cow moose and her calf feeding along the shoreline. They stuck tightly together for quite some time. Then the calf's attention wandered, while the mother slowly fed her way along until she was perhaps thirty or forty yards ahead. Suddenly discovering that he had been abandoned, the hulking calf, which certainly weighed well over two hundred pounds, began to squeak and whimper, much in the fashion of a distraught puppy. This whining continued until the calf had once again reached its mother's side. The incongruity of such pathetic little sounds emerging from an animal of that size made this incident truly amusing!

We also observed another interesting moose trait. Two cow moose emerged from the woods and begin to drink from a little rivulet. As we watched, they apparently decided that reaching down so far to drink was too difficult, so they both knelt on their front knees and continued to drink for some time. I have no idea whether or not moose commonly drink in this fashion, but they clearly do on occasion.

Although moose can become too numerous and cause serious problems for humans, they are nonetheless a welcome sight in reasonable numbers. Unique in appearance, towering in height, and massive in frame, they represent an evolutionary masterpiece, superbly designed to thrive in regions of deep snow and bitter cold. Their recent success assures us that we will continue to see and wonder at their majestic presence.

24

The All-American: The Bison

MYTHS

- The bison is a buffalo.
- Bison are still quite scarce.
- Bison are docile animals, safe to approach closely.

ONCE THEY WERE THE GLORY OF THE NORTH AMERICAN PRAIRIES, DARKENING THE PLAINS IN NUMBERS BEYOND COUNTING, AND SHAKING THE EARTH WITH THE IMPACT OF MILLIONS OF HOOVES. Then, in a few short decades, the seemingly inexhaustible herds of the majestic bison were reduced to such a pathetic remnant that extinction of this Plains monarch seemed imminent. The bison's near demise, followed by its renascence, is a tale worth repeating, filled with carelessness, ignorance, and greed, followed by concern and the beginnings of wisdom.

The North American bison is an astonishingly tough creature, perfectly adapted to an often incredibly harsh life. It's usually incorrectly referred to as a buffalo, but buffaloes are Old World animals such as the Cape buffalo of Africa and the water buffalo of India. The bison of North America is distinctly different, for it evolved with an ability to live in harmony with the North American prairies, a harmony so profound that bison and prairie became mutually dependent on each other. Because of this unique quality, it seems fitting to conclude this book with an account of a creature that epitomizes the Old West and the prairies of the United States and Canada.

Bison in North America are actually divided into two subspecies. The American, or Plains, bison is *Bison bison bison*. North of the prairies in south-

American (Plains) bison

ern Canada is the wood buffalo, *Bison bison athabaesca.* The two are quite similar, as evidenced by the fact that they are subspecies, rather than separate species. However, the wood bison is a bit longer than the Plains bison, as well as a little heavier in the hindquarters.

The story of the bison's near extinction is well known, but it bears repetition. Even relatively early European settlers had contact with bison, for bison ranged well eastward of the Great Plains into areas such as Indiana and Kentucky. There they lived in openings within partly forested areas. These were only the tip of the iceberg, however. As white settlers surged westward in the mid-1800s, especially after the American Civil War, they encountered seemingly endless herds of bison on the prairies.

These animals were the perfect denizens of the Great Plains. Moving in gigantic herds, they effectively gave the prairie an annual haircut and shave by cropping the plants almost to the ground. As the bison grazed, their dung—the famous buffalo chips—fertilized the land. This system permanently maintained a rich variety of annual and perennial plants, and held encroaching forestland at bay.

Simultaneously, the diversity of prairie plants nourished the bison and provided an ecosystem for which it was perfectly adapted. Able to tolerate drought and the searing heat of summer on the Plains, the bison was also able to withstand the shrieking, howling winds, driving snow, and subzero temperatures of prairie blizzards. With heads, necks, and front quarters insulated by incredibly dense, woolly fur, the bison simply faced into the blizzards and endured, where lesser creatures, such as domestic cattle, soon perished.

So huge were some of the bison herds that we can scarcely imagine them today. George Armstrong Custer wrote of leading troops through a herd for six consecutive days, during the last three of which the bison were steadily and continually moving across their path! These were part of a mighty population estimated as high as 75 million, although 60 million is the most widely accepted estimate.

For centuries the Plains Indians had shared the land with the bison and had developed an almost mystic relationship with the great, shaggy creatures. On foot for hundreds of years, and subsequently with the advent of horses obtained from Spanish explorers in the sixteenth century, these Native Americans hunted bison. Their prey provided them not only with food, but also with most of their other necessities: clothing and tepee covers from the hides; cups and ladles from the horns; knives, arrowheads, sled runners, and hoes from the bones; bowstrings from sinews; dried dung for fuel; and the skull and horns as ceremonial and religious objects.

Then, with railroads reaching their tentacles into the West, white settlers came in earnest, and the slaughter began. In one of the most disgraceful chapters in the pillage of North America's natural resources, this enormous population of bison was nearly extinguished within a few short years. With the aid of modern weaponry—repeating rifles and more powerful rifles with longer range—bison were hunted in every available manner. They were shot by men on foot, on horseback, and even leaning out the windows of trains.

Some of the killing was at least for the legitimate purpose of procuring meat and hides, but much was simply wanton slaughter. Bison were shot by the thousands, their tongues cut out as a delicacy, and the remainder left to rot. Even worse, thousands more were shot by wealthy "sportsmen" from train windows and wagons for no better purpose than the fun of it.

Exacerbating the situation was an underlying government policy of eliminating the bison as a means of solving the "Indian problem." Rid the Plains of bison, the reasoning went, and the Indians will vanish also. No less a figure than Lieutenant General Philip Sheridan, the great Union Civil War leader, declared in 1875, "These [hide hunters] have done in the last two years, and will do in the next year, more to settle the vexed Indian question than the entire Regular Army has done in the last thirty years. They are destroying the Indians' commissary . . . let them kill, skin, and sell until the buffaloes are exterminated. Then your prairies can be covered with speckled cattle and the festive cowboy. . . ." However misguided its aims, this policy was unquestionably effective in achieving its purpose.

Incredibly, by 1885 the bison were nearly gone. This creature, which had darkened the Plains in great, thundering rivers of living animals, and which was strong enough and fast enough to face down or outrun most predators, was no match for man and his rifles. By 1893 it was estimated that perhaps as few as three hundred bison remained.

Fortunately, a few ranchers and true sportsmen, appalled by what was happening, rounded up a few of the fast-dwindling remnant bison and began to raise them in captivity. In 1905 the noted biologist William Hornaday became the cofounder and first president of the American Bison Society, dedicated to saving and restoring the bison. Meanwhile, President Theodore Roosevelt successfully pushed Congress into establishing a number of wildlife preserves. Many of these, along with several national parks, were restocked with animals from private bison owners, and the former monarch of the Plains began its slow march back from the brink of extinction.

There seems to be a rather vague sense that the bison, though no longer endangered, is rather uncommon. In fact, it's thriving, and its numbers will eventually be limited only by available habitat. Today there are an estimated 250,000 bison in North America, found in nearly every province and state, including Alaska. Most are captive animals, but at least three herds aren't confined by fences.

Captive doesn't equate with tame, however. Bison often seem placid and docile, but they're not domestic livestock. Those most familiar with bison warn that they are *not* tame and should not be approached closely by anyone other than those experienced in handling them. Bison, they caution, despite their seeming docility, are dangerous because of their size, speed, and unpredictability.

A former National Wildlife Federation colleague of mine was an accidental participant in the sort of drama that bison experts warn against. He happened to be driving past a ranch where bison were confined by high and extremely heavy fencing. There he noticed a man on the outside taking pictures of a young boy within, standing beside a group of bison cows. As he took in this scene, he also noticed a buffalo bull some distance away, angrily pawing the ground in preparation for a charge.

Slamming on the brakes, my colleague jumped out of the car and climbed over the fence. There he seized the boy, tossed him over the fence, grabbed the top of the fence, and hoisted himself out of range just as the angry bison slammed into the fence just below his feet! His bravery and quick thinking had undoubtedly saved the lad's life.

This story has a rather dispiriting sequel. Rather than being grateful to my colleague and remorseful for his own stupidity, the father sued him because the son was slightly injured when he hit the ground. Fortunately justice prevailed: the judge promptly dismissed the lawsuit.

Three bison herds in the United States and one in Canada are essentially wild. That is, they're unfenced, although they're managed to one degree or another to control their numbers. One, called the Wild Bunch after Butch Cassidy's gang, roams freely in the wild fastnesses of the Henry Mountains in southern Utah (a portion of that herd also ranges near the northern rim of the Grand Canyon in northern Arizona). Another unfenced herd is located in Alaska. These two herds are kept in balance with their habitat by carefully regulated hunting. There is also a wild herd of wood bison in Wood Bison Park in Canada, the same park that's home to nesting whooping cranes.

Then there is the bison herd in Yellowstone National Park, a source of great controversy. Because national parks have a no-hunting policy, no control has been exerted over Yellowstone's bison. As might be expected, the result is overpopulation, and excess bison regularly spill outside the park's boundaries.

Because some of these bison are infected with brucellosis, a disease that causes domestic cows to abort their calves, Montana officials have killed the bison outside the park, ostensibly to protect ranchers' livestock. This seems reasonable on the surface, but there are some very troubling aspects to it. First, there has never been a documented case of brucellosis transmission from wild bison to domestic cattle. Second, Montana rejected a National Wildlife Federation proposal to reimburse nearby ranchers for vaccinating their livestock against brucellosis.

Third, a coalition of forty-six Native American Tribes, organized as the Intertribal Bison Cooperative, wants to relocate these Yellowstone escapees onto various tribal lands as a means of restoring at least a small portion of the tribes' natural and cultural heritage. Further, these transplanted animals would eventually generate surplus bison that could be used to restock various public lands. So far, Montana officials have chosen to kill these wandering bison rather than allow them to be moved to tribal lands.

Historically, wolves and Native Americans were the main predators of bison, and it may well be that the burgeoning population of wolves in Yellowstone Park will eventually assume that role. Currently the wolves seem to be concentrating their efforts mostly on elk, which are easier to kill than bison. As the park's overpopulation of elk is brought under control by the wolves, however, it seems likely that they'll begin to turn their attention to the bison.

In addition to these unfenced wild herds, there are also what I think can reasonably be termed semi-wild herds, which are fenced, but in such enormously large areas, encompassing tens of thousands of acres, that they can roam freely for all practical purposes. These include the bison in Custer State Park and Wind Cave National Park, among others.

Bison are not only incredibly hardy; they're also unrivaled for size in North America. Cows normally weigh about one thousand pounds, and bulls 1,300 to 1,500 pounds, but a really large bison bull can stand over six feet high at the shoulder and weigh over two thousand pounds! This is heavier than the largest moose, brown bear, or polar bear.

Bison are huge, but they're also fast and have immense endurance. They can run over thirty miles an hour at top speed, and at a slower pace they can run for hours without stopping. All in all, this makes the bison an exceptionally difficult target for even a very large predator.

The power and durability of the bison was highlighted by a bizarre contest that took place in 1907 between Pierre, a bison bull, and a series of Mexican fighting bulls. This incident began when several Mexican officials visited Fort Pierre, South Dakota, to view some bison confined on a ranch there. The officials weren't impressed by the slow movements and docile behavior of the bison, and made disparaging remarks about them. A Mexican fighting bull, they told their hosts, would make short work of a bison.

This prompted a retort, one thing led to another, and soon a fight was arranged in the bull ring in Juarez, Mexico. There, to settle what had escalated into a rather rancorous dispute involving the national pride of both countries, Pierre was to take on a succession of fighting bulls (assuming he outlasted the first one) in a quadruped version of a Demolition Derby.

When Pierre was brought into the ring, he placidly lay down in the warm sunshine, unfazed by the jeers of the hostile crowd. Then the Mexican fighting bull was brought in, accompanied by loud cheers.

After looking his adversary over, the Mexican bull finally approached Pierre, who got to his feet at that point, but made no other move. Four times the fighting bull attacked, and each time Pierre simply turned to meet his charge head-on.

A bison, with the bulk of its weight centered over its front quarters, pivots on its front legs, whereas a domestic bull pivots on its hind legs. The fighting bull was evidently baffled by its opponent's ability to pivot so quickly, and the results of its charges were stunning in more ways than one. The first time they collided, the Mexican bull staggered backward. The next time it dropped to its knees after the crash, and the third and fourth times it went all the way down. This was enough for the fighting bull, which broke off the engagement and sought a way out of the arena. At no time did Pierre make the slightest move to follow up his advantage. Like Ferdinand the Bull, he just wanted to be left in peace.

Two more fighting bulls were released in succession, with results virtually identical to those of the first encounter. When a fourth bull was brought on, however, Pierre finally became irritated; when the bull charged, Pierre responded by returning the charge. They met full-tilt in the middle of the ring with a horrendous crash, and the result was astounding. The Mexican bull shot backward as if it had been fired from a cannon, and landed in a heap. When it arose, it wanted no more of such a fearsome adversary and tried to get out of the ring as expeditiously as possible.

Although this contest was not perhaps the most politically correct mode of determining the relative strength of domestic bull and bison, it certainly provided a convincing demonstration of the latter's awesome power and toughness. Figuratively speaking, Pierre had defeated all comers with one hoof tied behind his back!

In keeping with their great size and strength, bison are long-lived. An average life span is twenty-five years, and cows can live to forty or more. Heifers sometimes breed when a year old, but most reach breeding age at two. Following a nine-month gestation, the cow usually gives birth to a single calf, though she may rarely have twins. As part of the bison's astonishing capacity for survival, a cow can successfully give birth to a calf and raise it in the midst of a raging, late-season blizzard. As might be expected of a grazing animal keeping pace with a constantly moving herd, the calves are quickly on their feet and able to follow their mothers.

There are many references to "buffalo wallows" in literature and historical accounts. Bison have short tails that aren't effective in keeping flies and other insects away from their heads and front quarters, so they enjoy rolling in dust or mud wallows. The latter, no doubt, also help to cool the bison in hot weather.

The bison's future is bright for a variety of reasons. Once raised mainly as a curiosity and a means of preserving the species, the bison is increasingly in demand as a meat animal. Bison meat is tender, flavorful, extremely nutritious, and very low in cholesterol. Consequently, more and more restaurants, specialty markets, and individuals are purchasing bison meat at a premium price.

Moreover, an increasing number of people are considering the substitution of bison for cattle on marginal rangelands. Some are private ranchers who see bison as a moneymaking alternative to cattle under harsh conditions. Others—private conservation organizations such as the Nature Conservancy and the Land Institute, or conservation-minded individuals like Ted Turner—seek to preserve and restore prairie ecosystems by restoring their vital natural component, the bison.

Beyond that, major national organizations, including the National Wildlife Federation, the National Audubon Society, and the Sierra Club, are promoting ways to protect and restore prairie lands, in part by restoring bison to some of their old haunts. There is also the 2,400-member National Bison Association, formed for the preservation, production, and marketing of bison, as well as the Wild Bison Foundation, dedicated to preserving our remaining free-roaming herds of bison.

Taken together, this is a formidable array of private interests and citizen power in various forms, representing a great upswing in support of additional bison restoration. We will not, of course, see the reappearance of bison in anything like their former numbers, for the prairie/bison ecosystem won't produce grain and meat in quantities sufficient for our needs. Nonetheless, we may well witness the return of relatively large herds of these great, shaggy, magnificent beasts—the quintessential North American mammal—grazing as in centuries past.

Resources

Some readers might wish to become more actively involved in the conservation of wildlife and the habitat on which it depends. To assist them, I've listed several sources of information, as well as some organizations that they might want to support.

The most comprehensive collection of useful information is the Conservation Directory, revised annually by the National Wildlife Federation. This can be purchased for $55 ($49.50 for NWF members and $44.00 for college students) by calling toll-free at (800) 477-5560, or ordered by fax at (540) 722-5399.

Although this price may seem high, the directory contains a wealth of information packed into its more than five hundred pages. Included are the congressional delegation from each state; federal agencies in the executive branch; international, national, and regional conservation organizations; government agencies and citizens' conservation groups within each state; Canadian government agencies; colleges and universities in the United States and Canada; and numerous other helpful resources.

I've also listed several national conservation/environmental organizations that readers may wish to support with volunteer activity, membership, or contributions. Some of these also have local or state chapters or affiliates.

I've selected these organizations based on four criteria. First, they are genuine conservation/environmental groups, concerned with the protection and enhancement of wildlife habitat, scientific management of wildlife resources, and the protection of rare, threatened, and endangered species. Second, they are respected mainstream organizations. Third, they base their work on scientific, rather than emotional, arguments. Fourth, they have a record of proven effectiveness in the wise use and proper protection of both wildlife and wildlife habitat.

There are, of course, numerous other worthwhile conservation organizations; I've merely listed some of the biggest and best. I would caution readers, however, to investigate other organizations very carefully before joining forces with them. Many groups with high-sounding names actually do little, if anything, to protect vital habitat or threatened and endangered species. The following organizations aren't in any particular order, except that the list is headed by what are generally acknowledged to be conservation's Big Three.

National Wildlife Federation, Conservation Education Center, 8925 Leesburg Pike, Vienna, VA 22184-0001; (703) 790-4000; fax: (703) 442-7332; Internet: http://www.nwf.org

National Audubon Society, 700 Broadway, New York, NY 10003-9501; (212) 979-3000

Sierra Club, Washington, DC, Office, 408 C St., NE, Washington, DC 20002; (202) 547-1141; fax (202) 547-6009; hotline (202) 675-2394

Wildlife Conservation Society, 185th St. and Southern Blvd., Bronx, NY 10460-1099; (718) 220-5100; fax (718) 220-7114

The Izaak Walton League of America, 707 Conservation Lane, Gaithersburg, MD 20878; (301) 548-0150

The Wilderness Society, 900 17th St., NW, Washington, DC 20006-2596; (202) 833-2300

Defenders of Wildlife, 1101 14th St., NW, Suite 1400, Washington, DC 20005; (202) 682-9400; fax (202) 682-1331; Internet: http://information@defenders.org; http://www.defenders.org

Selected Bibliography

Ball, John, et al. *National Audubon Society Field Guide to North American Birds: Eastern Region*. New York: Alfred A. Knopf, 1994.

Burt, William H., and Richard P. Grossenheider. *A Field Guide to the Mammals: North America North of Mexico*. Boston: Houghton Mifflin, 1998.

Conant, Roger, et al. *A Field Guide to Reptiles and Amphibians: Eastern and Central North America*. Boston: Houghton Mifflin, 1998.

Heinrich, Bernd. *Mind of the Raven: Investigations and Adventures with Wolf-Birds*. New York: Cliff Street Books (HarperCollins), 1999.

Henry, J. David. *How to Spot a Fox*. Shelburne, VT: Chapters Publishing, Ltd., 1993.

Henry, J. David. *Red Fox: The Catlike Canine*. Washington, D.C.: Smithsonian Institution Press, 1996.

Hunter, Malcolm L., Jr., John Albright, and Jane Arbuckle, eds. *The Amphibians and Reptiles of Maine*. Bulletin 838, Maine Agricultural Experiment Station, University of Maine, Orono 1992.

Mech, L. David, and Ian MacTaggart. *The Wolf: The Ecology and Behavior of an Endangered Species*. St. Paul, MN: University of Minnesota, 1985.

Mech, L. David. *The Way of the Wolf*. Stillwater, MN: Voyageur Press, 1991.

Mech, L. David, Michael K. Phillips, and Roger A. Caras. *The Arctic Wolf: Ten Years with the Pack*. Stillwater, MN: Voyageur Press, 1997.

Murie, Olaus J., and Roger Tory Peterson. *A Field Guide to Animal Tracks*. Boston: Houghton Mifflin, 1998.

Nelson, Richard. *Heart and Blood: Living with Deer in America*. New York: Alfred A. Knopf, 1997.

Peterson, Roger Tory, and Virginia Marie Peterson. *A Field Guide to the Birds: A Completely New Guide to All the Birds of Eastern and Central North America*. Boston: Houghton Mifflin, 1998.

Udvardy, Miklos D. F., and John Farrand, Jr. *National Audubon Society Field Guide to North American Birds: Western Region*. New York: Alfred A. Knopf, 1994.

Walker, Ernest Pillsbury. *Walker's Mammals of the World, Fifth Edition*. Revised by Ronald M. Nowak. Baltimore: Johns Hopkins University Press, 1991.

Index

M

N